The ESSENTIALS of

REGISTERED TRADEMARK

FLUID MECHANICS & DYNAMICS I

**Staff of Research and Education Association,
Dr. M. Fogiel, Director**

This book covers the usual course outline of Fluid Mechanics & Dynamics I. For more advanced topics, see *"THE ESSENTIALS OF FLUID MECHANICS & DYNAMICS II"*.

**Research and Education Association
61 Ethel Road West
Piscataway, New Jersey 08854**

THE ESSENTIALS OF
FLUID MECHANICS & DYNAMICS I ®

Printed in the United States of America

Library of Congress Catalog Card Number 87-61807

International Standard Book Number 0-87891-594-X

Revised Printing 1989

WHAT "THE ESSENTIALS" WILL DO FOR YOU

This book is a review and study guide. It is comprehensive and it is concise.

It helps in preparing for exams, in doing homework, and remains a handy reference source at all times.

It condenses the vast amount of detail characteristic of the subject matter and summarizes the **essentials** of the field.

It will thus save hours of study and preparation time.

The book provides quick access to the important facts, principles, theorems, concepts, and equations of the field.

Materials needed for exams, can be reviewed in summary form — eliminating the need to read and re-read many pages of textbook and class notes. The summaries will even tend to bring detail to mind that had been previously read or noted.

This "ESSENTIALS" book has been carefully prepared by educators and professionals and was subsequently reviewed by another group of editors to assure accuracy and maximum usefulness.

Dr. Max Fogiel
Program Director

CONTENTS

Chapter No.		**Page No.**
1	**INTRODUCTION**	1
1.1	Definition of a Fluid	1
1.2	Basic Laws in Fluid Mechanics	1
1.3	System and Control Volume	1
1.4	Dimensions and Units	2
2	**FUNDAMENTAL CONCEPTS**	5
2.1	Fluid as a Continuum	5
2.2	Fluid Properties	5
2.3	Two Viewpoints	6
2.4	One, Two, and Three Dimensional Flow	6
2.5	Pathlines, Streaklines, Streamlines, and Streamtubes	7
2.6	Stress Field	8
2.7	Newtonian Fluid Viscosity	9
2.8	Classification of Fluid Motion	11
2.9	Irrotational Flow	11
3	**FLUID STATICS**	13
3.1	The Basic Equation of Fluid Statics	13
3.2	The Standard Atmosphere	15
3.3	Absolute and Gage Pressures	16
3.4	The Simple Manometer	16

3.5   Hydrostatic Forces on Submerged Surfaces   17
3.6   Buoyancy and Stability   20
3.7   Fluids in Rigid-Body Motion   22

4   BASIC LAWS FOR SYSTEMS AND CONTROL VOLUMES   23

4.1   Basic Laws for a System   23
4.2   Reynolds Transport Equation   25
4.3   The Basic Laws Applied to the Control Volume   26

5   INTRODUCTION TO DIFFERENTIAL ANALYSIS OF FLUID MOTION   33

5.1   The Continuity Equation   33
5.2   The Differential Form of Momentum Equation   35
5.3   Euler's Equations   38
5.4   Bernoulli's Equation   39

6   CONSIDERATIONS FOR COMPRESSIBLE FLOW   41

6.1   Review of Thermodynamics   41
6.2   Propagation of Sound Waves   43
6.3   Local Isentropic Stagnation Properites   45
6.4   Critical Condition   45

7   ONE-DIMENSIONAL COMPRESSIBLE FLOW   47

7.1   Basic Equations for Isentropic Flow   47
7.2   Effects of Area Variation on Flow Properties in Isentropic Flow   49
7.3   Isentropic Flow of an Ideal Gas   50
7.4   Isentropic Flow in a Converging Nozzle   51
7.5   Adiabatic Flow in a Constant Area Duct with Friction   52
7.6   The Fanno Line   54
7.7   Frictionless Flow in a Constant Area Duct with Heat Addition   55

7.8 The Rayleigh Line 57
7.9 Normal Shock 58
7.10 Flow in a Converging-Diverging Nozzle 62
7.11 Oblique Shock 63

8 DIMENSIONAL ANALYSIS AND PHYSICAL SIMILARITY

67

8.1 Dimensional Analysis, Physical Similarity 67
8.2 Types of Physical Similarity 68
8.3 The Buckingham's Π Theorem 69
8.4 Dimensional Analysis of a Problem 69

CHAPTER 1

INTRODUCTION

1.1 DEFINITION OF A FLUID

A fluid is a substance that changes its shape continuously under the application of a shear stress no matter how slight the shear stress.

1.2 BASIC LAWS IN FLUID MECHANICS

Five basic laws are used for the solution of any problem in fluid mechanics. These basic laws are:

A) Conservation of mass

B) Newton's second law of motion

C) Conservation of momentum

D) The first law of thermodynamics

E) The second law of thermodynamics

1.3 SYSTEM AND CONTROL VOLUME

A) System

A system is a fixed, identifiable quantity of matter,

that may change in shape, position, and thermal condition. The system boundary separates the system from the surroundings (Fig. 1.1).

B) Control Volume

A control volume is a fixed, arbitrary volume in space through which fluid flows. The boundary of the control volume, is called the control surface (Fig. 1.2).

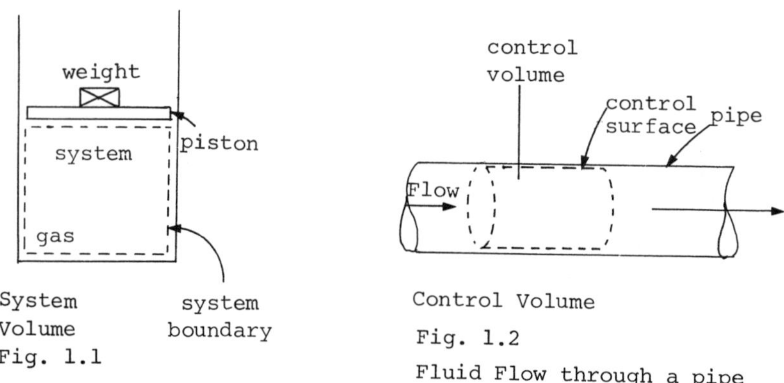

System
Volume
Fig. 1.1

system
boundary

Control Volume
Fig. 1.2
Fluid Flow through a pipe

1.4 DIMENSIONS AND UNITS

Dimensions are names given to measurable quantities, such as time, length, velocity, mass, volume, and force.

Primary dimensions are those dimensions which are independent of other dimensions.

Secondary dimensions are those dimensions which are expressed in terms of the primary dimensions.

A unit is a definite standard or measure of a dimension.

The four basic systems of units are shown in table 1.1. The physical equivalents between some of the units are listed in table 1.2.

Table 1.1   Systems of Units

	Dimensions	Units	Definition of a secondary unit
I	International System		
Primary Dimensions	Mass, M Length, L Time, t Temperature, T	Kilogram, Kg Meter, m Second, sec Kelvin, K	
Secondary Dimensions	Force, F	Newton, N	$1N = \dfrac{1\ Kg \cdot m}{sec^2}$
II	Absolute Metric System		
Primary Dimensions	Mass, M Length, L Time, t Temperature, T	Gram, gm Centimeter, cm Second, sec Kelvin, K	
Secondary Dimensions	Force, F	dyne	$1\ dyne = \dfrac{1\ gm \cdot cm}{sec^2}$
III	British Gravitational System		
Primary Dimensions	Force, F Length, L Time, t Temperature, T	Pound-force, lbf Foot, ft Second, sec Rankine, R	
Secondary Dimensions	Mass, M	slug	$1\ slug = \dfrac{1\ lbf \cdot sec^2}{ft}$
IV	English Engineering System		
Primary Dimensions	Force, F Mass, M Length, L Time, t Temperature, T	Pound-force, lbf Pound-mass, lbm Foot, ft Second, sec Rankine, R	
Secondary Dimensions			

Table 1.2

Equivalence Relations Between Units	
1 in. ≡ 2.54 cm	1 slug ≡ 32.2 lbm
1 ft. ≡ 30.5 cm	$1 \text{ g} \equiv 2.20 \times 10^{-3}$ lbm
1 ft ≡ 0.305 m	$1 \text{ g} \equiv 0.685 \times 10^{-4}$ slug
5,280 ft ≡ 1 mile	1 ft-lb ≡ 0.001285 Btu
1 gal (U.S.) ≡ 0.1337 ft^3	$1 \text{ joule} \equiv 9.48 \times 10^{-4}$ Btu
1 lbf ≡ 16 oz	1 hp ≡ 550 ft-lb/sec
1 newton ≡ 10^5 dynes	1°K ≡ 1.8°R (where °K =
1 lbf ≡ 445,000 dynes	°C + 273° and °R
	= °F + 460°)

NOTES

In the solution of problems, the following logical steps should be taken:

A) State the information given.

B) State the unknowns to be found.

C) Select and draw the system or control volume to be used. Be sure to label the boundaries of the system or control volume, and use appropriate coordinate directions.

D) Write the basic equations and laws that are necessary to solve the problem.

E) List the appropriate assumptions.

F) Using a set of units, substitute numerical values into equations or laws.

G) Check the answer to make sure that it is reasonable with the assumptions made.

CHAPTER 2

FUNDAMENTAL CONCEPTS

2.1 FLUID AS A CONTINUUM

When a fluid is treated as an infinitely divisible substance, one which is composed of many molecules in constant motion and in collision with each other, it is said to be in continuum.

2.2 FLUID PROPERTIES

As a consequence of the continuum assumption, the fluid's properties are considered to be continuous functions of position and time. Thus, the representation of fluid properties are:

A) Density field:

The density field is represented by

$$\rho = \rho(x,y,z,t) \tag{2.1}$$

where x, y, z are the space coordinates at any instant of time.

B) Velocity field:

The velocity field is represented by

$$\vec{v} = \vec{v}(x,y,z,t) \tag{2.2}$$

If the properties and flow characteristics at each position in space remain invariant with time, the flow is called steady flow.

For steady flow

$$\rho = \rho(x,y,z) \quad \text{or,} \quad \frac{\partial \rho}{\partial t} = 0 \qquad (2.3)$$

and

$$\vec{v} = \vec{v}(x,y,z) \quad \text{or,} \quad \frac{\partial \vec{v}}{\partial t} = 0 \qquad (2.4)$$

A time dependent flow, on the other hand, is designated as an unsteady flow.

Uniform flow is a flow in which the magnitude and direction of the velocity vector do not change with time.

2.3 TWO VIEWPOINTS

There are two possibilities to describe the velocity field in computations involving the motion of fluid particles. The velocity of particles at any time can be expressed using fixed coordinates x, y, z in the flow field. Mathematically it may be given by $\vec{v} = \vec{v}(x,y,z,t)$. This procedure is called the Eulerian viewpoint.

Now, to study any one particle in the flow, one must follow the particle as it moves. This is obtained by using the corresponding time functions for each particle. Mathematically it may be expressed by $\vec{v} = \vec{v}[x(t),y(t),z(t)]$, this procedure is called the Lagrangian viewpoint.

2.4 ONE, TWO, AND THREE DIMENSIONAL FLOW

A one-dimensional flow is a flow in which the parameters are expressed as functions of one space coordinate and time.

A two-dimensional flow is a flow in which the parameters are expressed as functions of two space coordinates and time.

A three-dimensional flow is a flow in which the parameters are expressed as functions of three space coordinates and time.

2.5 PATHLINES, STREAKLINES, STREAMLINES, AND STREAMTUBES

Pathline is a line traced out by a moving particle in time.

Streakline is a line traced out by all fluid particles which at some time passed through one fixed location in space.

Streamlines are imaginary lines drawn in the flow whose tangents are parallel to the velocity vector at a given instant of time.

Streamtube is an elementary region of the flow whose walls are made up of streamlines (Fig. 2.1).

In the case of steady flow $\left(\frac{\partial v}{\partial t} = 0 \right)$, pathlines, streaklines, and streamlines all coincide.

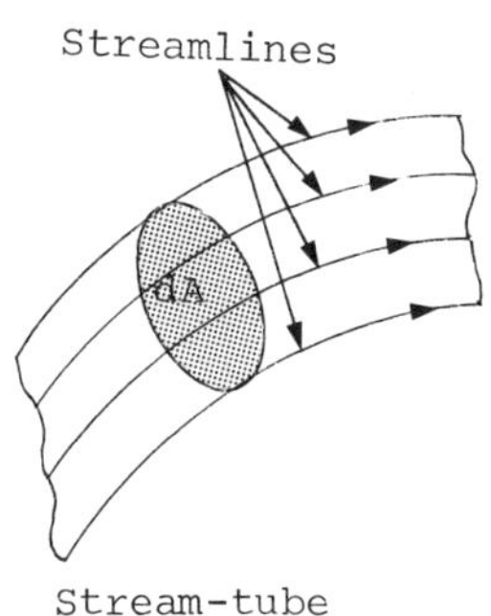

Fig. 2.1

2.6 STRESS FIELD

A) Surface and Body Forces

Surface forces are all of the forces that act upon the boundaries of a medium by direct contact.

Forces which develop without physical contact and are distributed over a volume of the fluid, are called body forces. Two examples are gravitational and electromagnetic forces.

B) Stress at a Point

The stress at a point is defined as

$$\text{stress} = \lim_{\delta\vec{A}\to 0} \frac{\delta\vec{F}}{\delta\vec{A}} \tag{2.5}$$

where

$\delta\vec{F} \equiv$ The force acting at the point B, as shown in Fig. 2.2.

$\delta\vec{A} \equiv$ The area element.

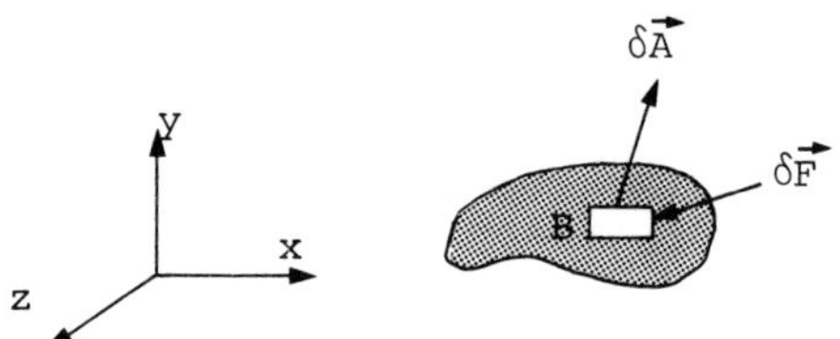

Definition of stress

Fig. 2.2

We can write the definition of stress in terms of components as

$$T_{ij} = \lim_{\delta A_i \to 0} \frac{\delta F_j}{\delta A_i} \tag{2.6}$$

where T_{ij} denotes the stress acting on an i plane in the j direction and where i and j each can stand for x, y, or z. For example, $T_{xz} = \lim\limits_{\delta A_x \to 0} \dfrac{\delta F_z}{\delta A_x}$ defines the stress on an x plane in the y direction (Fig. 2.3).

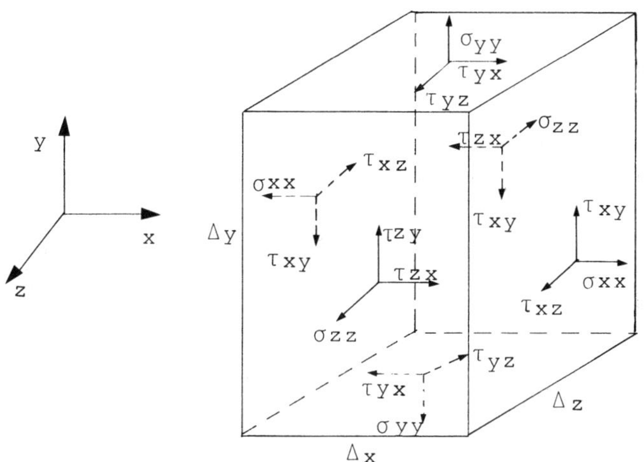

Fig. 2.3   Diagram showing stresses

The stress at a point is specified by the nine compoments

$$\begin{bmatrix} \sigma_{xx} & \tau_{xy} & \tau_{xz} \\ \tau_{yx} & \sigma_{yy} & \tau_{yz} \\ \tau_{zx} & \tau_{zy} & \sigma_{zz} \end{bmatrix} \qquad (2.7)$$

where σ has been used to denote a normal stress and τ to denote a shear stress.

2.7 NEWTONIAN FLUID VISCOSITY

Fluids may be classified according to the relation between the applied shear stress and the rate of deformation of the fluid. Fluids in which the shear stress is proportional to the rate of deformation are called Newtonian fluids (Fig. 2.5), or

$$\tau_{yx} \; \alpha \; \frac{du}{dy} \tag{2.8}$$

where

$$u \equiv \text{The velocity of the upper plate (Fig. 2.4)}$$

$$\frac{du}{dy} \equiv \text{Rate of deformation}$$

$$\tau_{yx} \equiv \text{The shear stress}$$

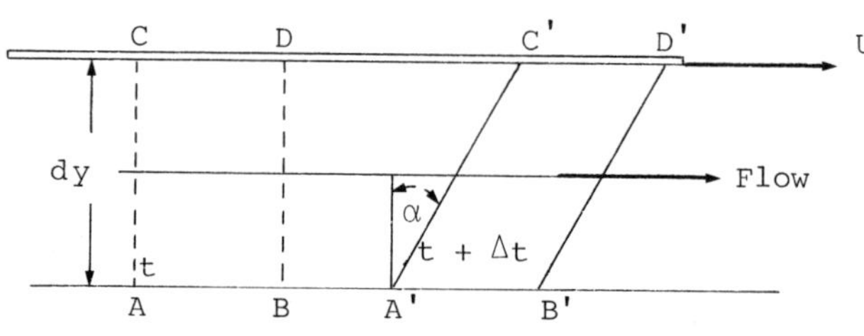

Fig. 2.4

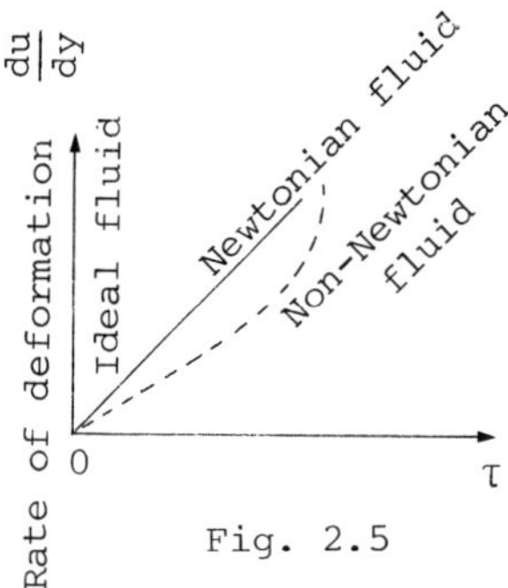

Fig. 2.5

The constant of proportionality in equation 2.8 is the coefficient of viscosity μ. Thus, equation 2.8 may be written as

$$\tau_{yx} = \mu \frac{du}{dy} \tag{2.9}$$

which is the Newton's law of viscosity.

The ratio of the absolute viscosity, μ, to the density, ρ, of a fluid is called kinematic viscosity and is represented by the symbol ν,

therefore,
$$\nu = \frac{\mu}{\rho} \tag{2.10}$$

2.8 CLASSIFICATION OF FLUID MOTION

One possible classification of fluid mechanics on the basis of observable physical characteristics of flow fields is shown in Fig. 2.6

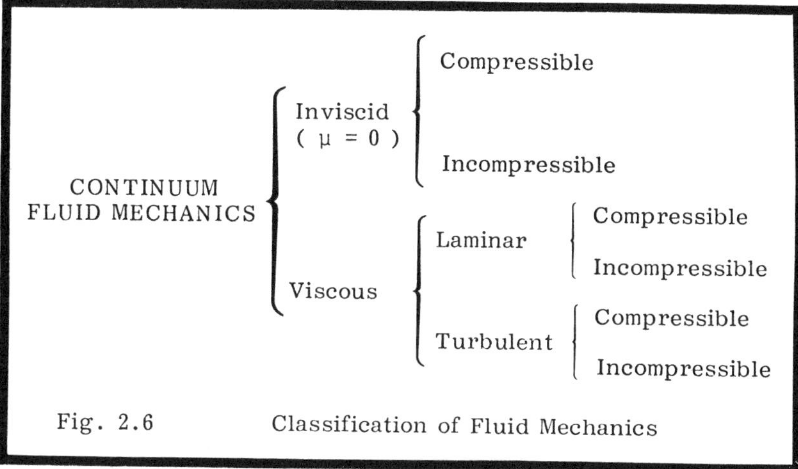

Fig. 2.6   Classification of Fluid Mechanics

A) Laminar Flow is a flow pattern characterized by motion in layers or laminae.

B) Turbulent Flow is a flow pattern characterized by random movements of fluid particles in all directions.

C) Incompressible Flow is a flow in which variations in density can be assumed to be constant. (ρ = const.)

D) Compressible Flow is a flow in which variations in density cannot be neglected. ($\rho \neq$ const.)

E) Inviscid fluids are fluids in which the fluid viscosity, μ, is assumed to be zero, ($\mu = 0$)

F) Viscous fluids are fluids in which the fluid viscosity, μ, cannot be neglected, ($\mu \neq 0$)

2.9 IRROTATIONAL FLOW

For a fluid element, the angular velocity, $\vec{w}$, is given by the relation

$$\boxed{\vec{w} = \frac{1}{2} \nabla \times \vec{v}}$$ (2.11)

or

$$\vec{w} = \frac{1}{2}\left[\left(\frac{\partial w}{\partial y} - \frac{\partial v}{\partial z}\right)\vec{i} + \left(\frac{\partial u}{\partial z} - \frac{\partial w}{\partial x}\right)\vec{j} + \left(\frac{\partial v}{\partial x} - \frac{\partial u}{\partial y}\right)\vec{k}\right]$$ (2.12)

where,

$$\boxed{\nabla \times \vec{v} = \operatorname{curl} \vec{v}}$$

$$\nabla \stackrel{\Delta}{=} \text{vector operator del} = \hat{i}\frac{\partial}{\partial x} + \hat{j}\frac{\partial}{\partial y} + \hat{k}\frac{\partial}{\partial z}$$

$$\vec{v} = u\hat{i} + v\hat{j} + w\hat{k}$$

u,v,w = The velocity compoments in the x,y,z directions, respectively.

When the angular velocity vector, $\vec{w}$, is equal to zero, the flow is called irrotational or,

$$\boxed{\vec{w} = \frac{1}{2}\nabla \times \vec{v} = 0}$$ (2.13)

or

$$\frac{\partial w}{\partial y} - \frac{\partial v}{\partial z} = \frac{\partial u}{\partial z} - \frac{\partial w}{\partial \cdot x} = \frac{\partial v}{\partial x} - \frac{\partial u}{\partial y} = 0$$ (2.14)

or,

$$\frac{\partial rV_\theta}{\partial r} - \frac{\partial V_r}{\partial \theta} = 0 \quad \text{(polar coordinate)}$$ (2.15)

The vorticity vector ($\vec{\zeta}$) for a fluid element is defined as:

$$\boxed{\vec{\zeta} = 2\vec{w} = \nabla \times \vec{v}}$$ (2.16)

In cylindrical coordinates the vorticity vector is:

$$\boxed{\begin{aligned}
\nabla \times \vec{v} &= \hat{i}_r\left(\frac{1}{r}\frac{\partial V_z}{\partial \theta} - \frac{\partial V_\theta}{\partial z}\right) + \hat{i}_\theta\left(\frac{\partial v_r}{\partial z} - \frac{\partial v_z}{\partial r}\right) \\
&\quad + \hat{i}_z\left(\frac{1}{r}\frac{\partial rV_\theta}{\partial r} - \frac{1}{r}\frac{\partial v_z}{\partial \theta}\right) \\
\text{where} \quad \vec{V} &= \hat{i}_r V_r + \hat{i}_\theta v_\theta + \hat{i}_z v_z
\end{aligned}}$$

CHAPTER 3

FLUID STATICS

3.1 THE BASIC EQUATION OF FLUID STATICS

To determine the pressure field within the fluid a differential element shown in Figure 3.1 is used.

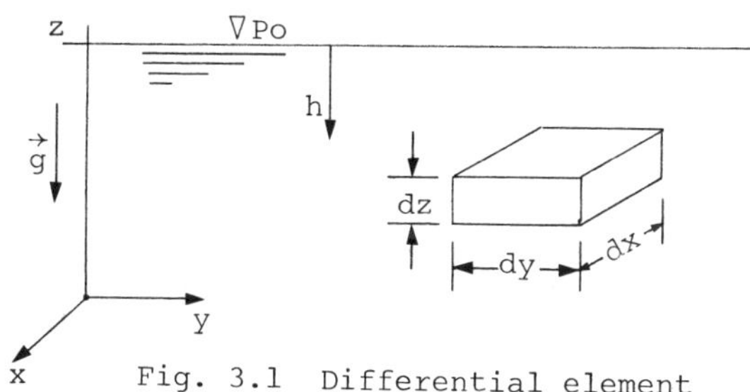

Fig. 3.1 Differential element

For this differential fluid element the following equation applies:

$$-\text{grad } p + \rho\vec{g} = 0 \tag{3.1}$$

where:

$$\text{grad } p \equiv \nabla p \equiv \left(\hat{i}\,\frac{\partial p}{\partial x} + \hat{j}\,\frac{\partial p}{\partial y} + \hat{k}\,\frac{\partial p}{\partial z}\right)$$

$$\equiv \left(\hat{i}\,\frac{\partial}{\partial x} + \hat{j}\,\frac{\partial}{\partial y} + \hat{k}\,\frac{\partial}{\partial z}\right) p \tag{3.2}$$

The term in parentheses is called the gradient of the pressure.

The physical significance of the ∇p is the pressure force per unit volume at a point.

$\rho \vec{g} \equiv$ body force per unit volume

$\rho \equiv$ density

$\vec{g} \equiv$ the local gravity vector

By expanding equation (3.1) into components, we find

$$x\text{-direction} \qquad -\frac{\partial p}{\partial x} + \rho g_x = 0 \qquad\qquad (3.3)$$

$$y\text{-direction} \qquad -\frac{\partial p}{\partial y} + \rho g_y = 0 \qquad\qquad (3.4)$$

$$z\text{-direction} \qquad -\frac{\partial p}{\partial z} + \rho g_z = 0 \qquad\qquad (3.5)$$

If the coordinate system is chosen such that the z-axis is directed vertically, then

$$g_x = 0, \quad g_y = 0, \quad \text{and} \quad g_z = -g$$

Because the pressure P varies only in the z direction and because it is not a function of x and y, we may use a total derivative in equation (3.5). Then the equation (3.5) becomes

$$\frac{dp}{dz} = -\rho g = -\gamma \qquad\qquad (3.6)$$

where γ = specific weight

This equation is the basic pressure-height relation of fluid statics. It holds under the following restrictions:

1) Static fluid

2) Gravity is the only body force

3) The z-axis is vertical

A) Incompressible Fluid

In an equation for an incompressible fluid, ρ = constant (Fig. 3.1). Therefore, for constant gravity, the equation (3.6) becomes

$$p = p_0 + \rho gh \qquad (3.7)$$

where

$p_0 \equiv$ the pressure p at the reference level

$h \equiv z-z_0$ (h measured positive downward)

$\rho \equiv$ density

B) Compressible Fluid

If we know the manner in which the specific weight, γ, varies, the pressure variation in a compressible fluid can be evaluated using equation (3.6).

3.2 THE STANDARD ATMOSPHERE

The internationally accepted properties of the standard atmosphere are summarized in Table 3.1:

Table 3.1 U.S. Standard Atmosphere (Conditions here are at sea level)		
Property	English Units	SI Units
Temperature, T	59°F	288 K
Pressure, p	14.696 psia	101.3k Pa(abs)
Viscosity, μ	$3.719 \cdot 10^{-7}$ lbf $\cdot$sec/ft^2	$1.781 \cdot 10^{-5}$ kg/m sec
Density, ρ	0.002377 slug/ft^3	1.225 kg/m^3
Specific weight, γ	0.07651 lbf/ft^3	–

The temperature is assumed to decrease linearly with height according to the relation,

$$T = (519 - 0.00357z)\ {}^\circ R \qquad (3.8)$$

where

$z \equiv$ The elevation above sea level in feet.

3.3 ABSOLUTE AND GAGE PRESSURES

Pressure values are stated with respect to a reference pressure level. If the reference pressure level is the vacuum then the pressure is called absolute. The difference between the absolute pressure and the surrounding pressure (mainly atmospheric pressure) is called gage pressure

$$P_{gage} = P_{absolute} - P_{atm} \tag{3.9}$$

or,

$$P_{absolute} = P_{atm} + P_{gage}$$

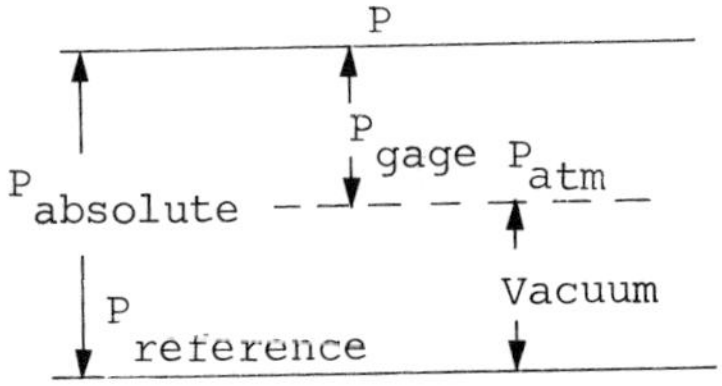

Fig. 3.2 Absolute and gage pressure

3.4 THE SIMPLE MANOMETER

Manometers are instruments that use liquid to determine differences in pressure between two points, or between a certain point and the atmosphere.

A manometer consisting of n different substances with different specific gravities $\gamma_1, \gamma_2, \ldots, \gamma_n$ is shown in Fig. 3.3.

To determine the pressure difference, $(P_A - P_n)$, (Fig. 3.3), the following equation is used.

$$(P_A - P_n) = (P_A - P_{A_1}) + (P_{A_1} - P_{A_2}) + (P_{A_2} - P_{A_3}) + (P_{A_3} - P_{A_4}) +$$

$$+\ldots+(P_{An-1}-P_n)$$

$$(3.10)$$

$$= -\gamma_1 z_1 - \gamma_2 z_2 + \gamma_3 z_3 + \gamma_4 z_4 \ldots -\gamma_n z_n$$

where

$$\gamma_1,\ \gamma_2,\ \gamma_3,\ \gamma_4,\ \ldots,\ \gamma_n = \text{the specific gravities of the substances}$$

$$z_1,\ z_2,\ z_3,\ \ldots,z_n = \text{the distances between two successive points in the columns of the manometer.}$$

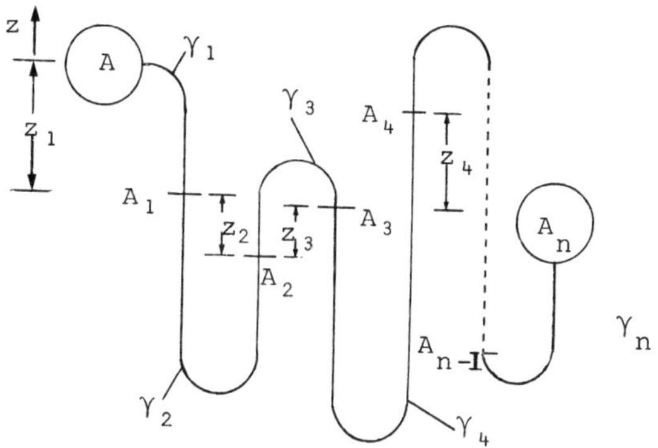

Fig. 3.3 Simple manometer

Note: The distances z_1, z_2, z_3,...,z_n are considered as positive if the end point, A_3, is at a higher position than the start point, A_2, (Distance z_3), or as negative if the end points, A_1, is at lower position than the start point, A (Distance z_1).

3.5 HYDROSTATIC FORCES ON SUBMERGED SURFACES

In order to determine the force acting on a submerged surface we must specify the magnitude, the direction and the line of action of the resultant force ($\vec{F}_R$).

A) Hydrostatic Force on a Plane Surface Submerged in a Static Incompressible Fluid

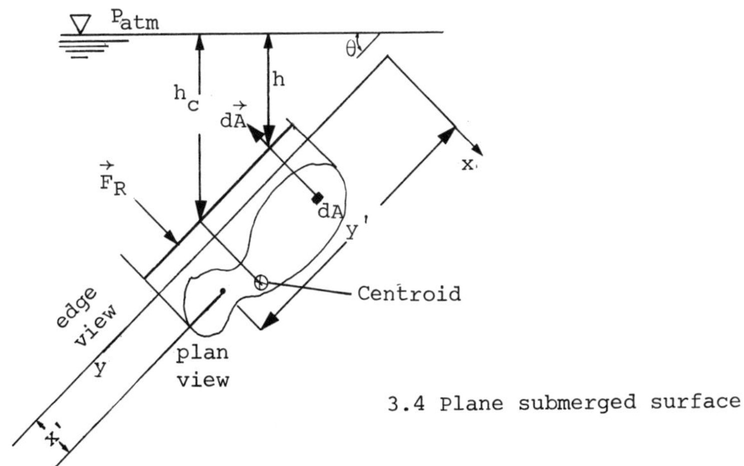

3.4 Plane submerged surface

The magnitude of the resultant force acting on the submerged suface is

$$F_R = \left| \vec{F}_R \right| = \int p\,dA$$

where A is the surface area or,

$$\boxed{F_R = \gamma h_c A} \qquad (3.11)$$

The direction of $\vec{F}_R$ is normal to the surface and the line of action passes through the points x', y', which can be located by

$$y' = y_c + \frac{I_c}{A_{yc}} \qquad \frac{I_c}{A_{yc}} > 0 \qquad (3.12)$$

$$x' = x_c + \frac{(I_{xy})_c}{A_{yc}} \qquad (3.13)$$

where I_c, $(I_{xy})_c$ is the moment of inertia about its center of gravity axes, and x_c, y_c is the center of gravity coordinates.

B) Hydrostatic Force on Curved Submerged Surfaces

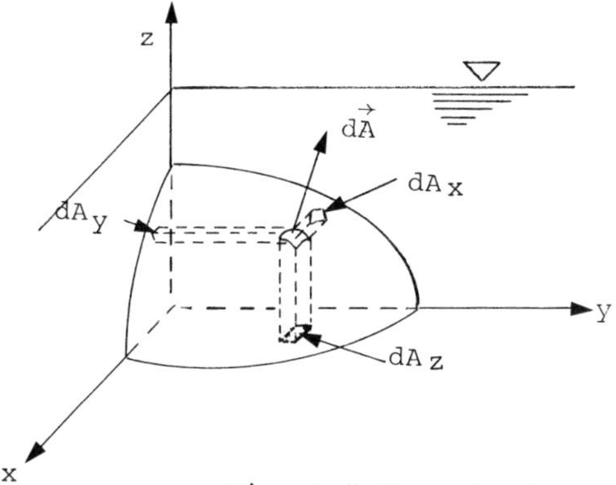

Fig. 3.5 Curved submerged surface

The methods used on plane surfaces can be used to determine the force on curve submerged surfaces.

For the submerged curved surface of Figure 3.5 the pressure force acting on the element of area $\vec{dA}$ is

$$\vec{F}_R = \hat{i}\, F_{Rx} + \hat{j}\, F_{Ry} + \hat{k}\, F_{Rz} = -\int_A p\, \vec{dA} \qquad (3.14)$$

where

$$p = p_0 + \rho g h$$

The components of the force in the x and y direction are

$$F_{Rx} = \pm \int_{Ax} p\, dA_x \qquad (3.15)$$

and

$$F_{Ry} = \pm \int_{Ay} p\, dA_y \qquad (3.16)$$

where $dA_x = dA\cos\theta_x$, and is the projection of the area element dA on a plane perpendicular to the x-axis.

$dA_y = dA\cos\theta_y$, and is the projection of the area element dA on a plane perpendicular to the y-axis.

The component of the force in the z-direction is

$$F_z = - \int_{A_z} \int_{z'}^{z_0} \rho g \, dz \, dA_z \qquad (3.17)$$

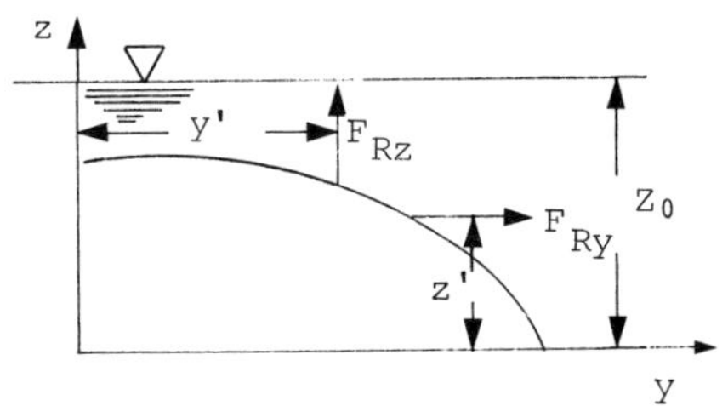

Fig.3.6 Two-dimensional curved submerged surface.

The lines of action of the components for a curved submerged surface are given by

$$z' = \frac{1}{F_{Ry}} \int_{A_y} z p \, dA_y \qquad (3.18)$$

$$y' = - \frac{1}{F_{Rz}} \int_{A_z} y p \, dA_z \qquad (3.19)$$

3.6 BUOYANCY AND STABILITY

A) Buoyancy of a Body

The resultant vertical force exerted on a body by a static fluid in which it is submerged or floating, is called the buoyant force.

The magnitude of this force is

$$\boxed{F_z = \int \rho g(z_2 - z_1)\,dA = \rho g \,\rlap{/}V} \qquad\qquad (3.20)$$

where ρ = density of the fluid

g = gravitational constant

$\rlap{/}V$ = volume of the body

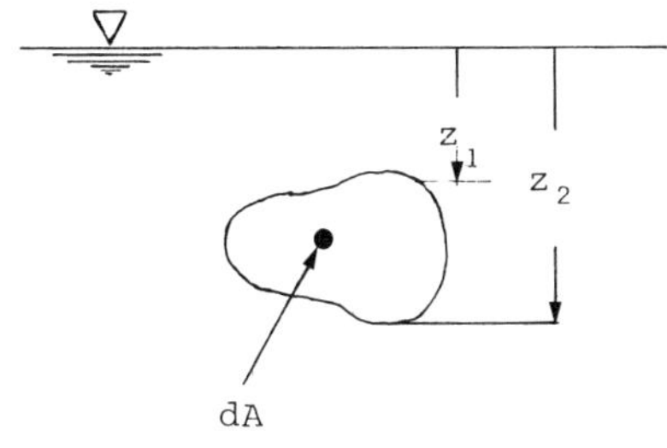

Fig. 3.7

The buoyant force for an incompressible fluid goes through the centroid of the volume displaced by the body.

B) Stability

For a completely submerged body, the condition for stability (stable equilibrium) is that the center of gravity, G, must be directly below the center of buoyancy.

The vertical alignment of B and G is important for stability. If G and B coincide, neutral equilibrium is obtained.

In a floating body (Fig. 3.8) stable equilibrium can be achieved even when G is above B. The magnitude of the length GA serves as a measure of the stability of a floating body.

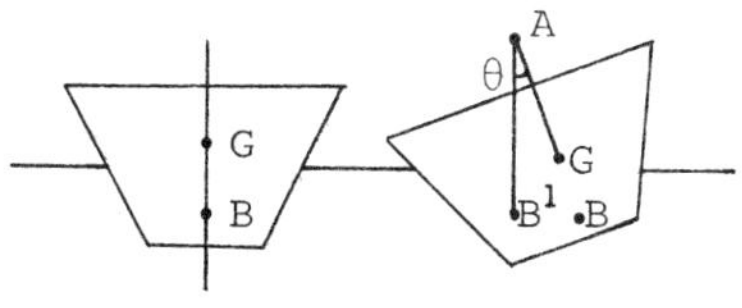

Fig. 3·8

Note: When a body undergoes a displacement and then returns to its original position, the body is in stable equilibrium. If the body does not return to its original position, but moves further away from it, the body is in unstable equilibrium. If the body adopts a new position the body is in neutral equilibrium.

3.7 FLUIDS IN RIGID-BODY MOTION

A) Accelerating Fluid

The equation of motion for a uniformly accelerating fluid is given by

$$-\text{grad } p + \rho \vec{g} = \rho \vec{a} \qquad (3.21)$$

where

$$\text{grad } p = \text{pressure force per unit volume}$$
$$\rho \vec{g} = \text{body force per unit volume at a point}$$
$$\rho = \text{density}$$
$$\vec{g} = \text{gravitational constant}$$
$$\vec{a} = \text{acceleration of the fluid}$$

The vector equation (3.20) has three components which are

$$\frac{\partial p}{\partial x} + \rho g_x = \rho a_x \qquad \text{(x-direction)} \qquad (3.22)$$

$$\frac{\partial p}{\partial y} + \rho g_y = \rho a_y \qquad \text{(y-direction)} \qquad (3.23)$$

$$\frac{\partial p}{\partial z} + \rho g_z = \rho a_z \qquad \text{(z-direction)} \qquad (3.24)$$

B) Rotating Fluid

For a fluid which rotates with a constant angular velocity, ω, about an axis z, equation (3.21) is reduced to

$$-\left.\frac{\partial p}{\partial z}\right|_z + \rho g_z = 0 \qquad (3.25)$$

CHAPTER 4

BASIC LAWS FOR SYSTEMS AND CONTROL VOLUMES

4.1 BASIC LAWS FOR A SYSTEM

A) Newton's Second Law

This law states that the total external force acting upon a system which is moving relative to an inertial reference is equal to the rate of change of the linear momentum of the system. Thus,

$$\boxed{\vec{F} = \vec{F}_S + \vec{F}_B = \frac{d\vec{P}}{dt}}$$

(4.1)

where

$$\vec{F}_S = \int_{CS} \vec{T}d\vec{A} = \text{Surface forces acting on the control volume.}$$

$$\vec{F}_B = \int_{CV} \vec{B}pd\Psi = \text{Body forces acting inside the control volume}$$

$$\vec{P} = \text{Linear momentum of the system}$$

$$= \int_{\Psi} V\rho d\Psi$$

$$V = \text{Velocity of the system}$$

$$\rho = \text{Density}$$

$$\mathcal{V} = \text{Volume}$$

B) Conservation of Mass

The conservation of mass simply states that the mass M of a system is always constant, or

$$\left.\frac{dM}{dt}\right|_{system} = 0 \qquad (4.2)$$

where

$$M = \int_{M,system} dm = \int_{\mathcal{V},system} \rho\,d\mathcal{V}$$

C) Moment of Momentum

This law states that the total torque acting on the system is equal to the rate of change of angular momentum, or

$$\vec{T} = \frac{d\vec{H}}{dt} \qquad (4.3)$$

where

$$\vec{H} = \text{The angular momentum of the system}$$

$$= \int_M \vec{r} \times \vec{V}dm = \int_{\mathcal{V}} \vec{r} \times \vec{V}d\mathcal{V}$$

$$\vec{T} = \text{The total torque of the system}$$

$$= \vec{r} \times \vec{F}_S + \int_M \vec{r} \times \vec{g}dm + \vec{T}_{shaft}$$

D) The First Law of Thermodynamics

This law states that

$$\delta Q + \delta W = dE \qquad (4.4)$$

where

$$Q = \text{Heat transferred to the system}$$

$$W = \text{Work done on the system}$$

$$E = \text{Total energy of the system}$$

$$= \int_M e\, dm = \int_V e\rho\, d\!\!\!\!\;\text{V}$$

$$e = u + \frac{V^2}{2} + gz$$

In rate form, equation (4.4) can be written as:

$$\boxed{\dot{Q} + \dot{W} = \left.\frac{dE}{dt}\right|_{\text{system}}} \qquad (4.5)$$

E)   The Second Law of Thermodynamics

This law states that

$$dS \geq \frac{\delta Q}{T} \qquad (4.6)$$

where

$$\delta Q = \text{Heat transferred to a system}$$

$$T = \text{Temperature}$$

$$dS = \text{Change in entropy, } S, \text{ of the system}$$

On a rate basis, equation (4.6) can be written as

$$\boxed{\left.\frac{dS}{dt}\right|_{\text{system}} \geq \frac{\dot{Q}}{T}} \qquad (4.7)$$

where

$$S_{\text{system}} = \int_M s\, dm = \int_V s\rho\, d\!\!\!\!\;\text{V}$$

4.2 REYNOLDS TRANSPORT EQUATION

A general relation between the rate of change of any arbitrary extensive property, N, of a system and the time variations of this property associated with the control volume is given by

$$\left.\frac{dN}{dt}\right|_{\text{system}} = \frac{\partial}{\partial t} \int_{CV} n\rho \, d\forall + \int_{CS} n\rho \vec{V} d\vec{A} \qquad (4.8)$$

$\left.\dfrac{dN}{dt}\right|_{\text{system}} \equiv$ Total rate of change of any extensive property of the system

$\dfrac{\partial}{\partial t} \displaystyle\int_{CV} n\rho d\forall \quad$ The time rate of change of the extensive property N in the control volume

$n = N$ per unit mass (Intensive property)

$\rho \, d\forall =$ An element of mass in the control volume

$\displaystyle\int_{CV} n\rho d\forall =$ The total amount of the extensive property N in the control volume

$\displaystyle\int_{CS} n\rho \vec{V} d\vec{A} \equiv$ Net rate of efflux of the extensive property N, crossing the control surface

$\rho \vec{V} d\vec{A} \equiv$ The rate of mass efflux crossing the element of area $d\vec{A}$ per unit of time

$n\rho \vec{V} d\vec{A} \equiv$ The rate of efflux of the extensive property N crossing $d\vec{A}$

4.3 THE BASIC LAWS APPLIED TO THE CONTROL VOLUME

A) Continuity Equation

The control volume formulation of the conservation of mass is given by the relation

$$\frac{\partial}{\partial t} \int_{CV} \rho \, d\forall + \int_{CS} \rho \vec{V} d\vec{A} = 0 \qquad (4.9)$$

where

$$\frac{\partial}{\partial t} \int_{CV} \rho \, dV = \text{The rate of change of mass within the control volume}$$

$$\int_{CS} \rho \vec{V} d\vec{A} = \text{The net rate of mass efflux through the control surface}$$

For steady or unsteady incompressible flow, the continuity equation is reduced to

$$\boxed{\int_{CS} V d\vec{A} = 0}$$
(4.10)

or,
$$\boxed{Q = A_1 V_1 = A_2 V_2}$$

where

$$Q = \text{Volumetric flow rate}$$

$$A_1, A_2 = \text{Cross-sectional areas at the entrance and exit of the control volume, respectively}$$

$$V_i, V_2 = \text{Fluid velocities at the entrance and exit of the control volume, respectively}$$

For any steady flow, the continuity equation is reduced to

$$\boxed{\int_{CV} \rho \vec{V} d\vec{A} = 0}$$
(4.11)

B) Linear Momentum Equation for the Inertial Control Volume

For an inertial control volume the linear momentum equation is given by the relation

$$\boxed{\vec{F} = \vec{F}_S + \vec{F}_B = \int_{CS} \vec{V} \rho \vec{V} dA + \frac{\partial}{\partial t} \int_{CV} \vec{V} \rho \, dV}$$
(4.12)

The vector form of the equation (4.12) may be written as

$$\boxed{F_i = \int_{CS} T_i dA + \int_{CV} B_i \rho \, dV = \int_{CS} V_i \rho \vec{V} d\vec{A} + \frac{\partial}{\partial t} \int_{CV} V_i \rho \, dV}$$
(14.13)

for $\quad i = x, y, z$

Note: The sign of the scalar quantity $\rho \vec{V} d\vec{A}$ depends on the dirction of flow. $\rho \vec{V} d\vec{A}$ is positive where flow is out through the control volume, and negative where flow is in through the control volume.

C) Linear Momentum Equation for a Control Volume Moving with Constant Velocity

This equation can be written as

$$\vec{F} = \vec{F}_S + \vec{F}_B = \int_{CS} \vec{V}_{xyz}\, \rho \vec{V}_{xyz}\, d\vec{A} + \frac{\partial}{\partial t} \int_{CV} \vec{V}_{xyz}\, \rho\, d\forall \qquad (4.14)$$

where the xyz subscript shows that the velocities must be measured relative to a reference coordinate system xyz (see Fig. 4.1).

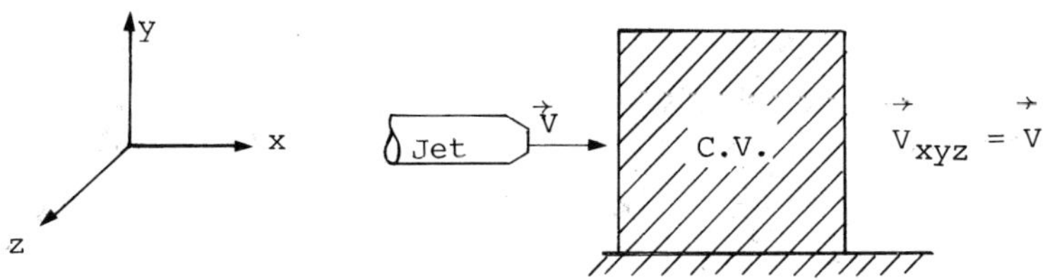

Stationary Control Volume

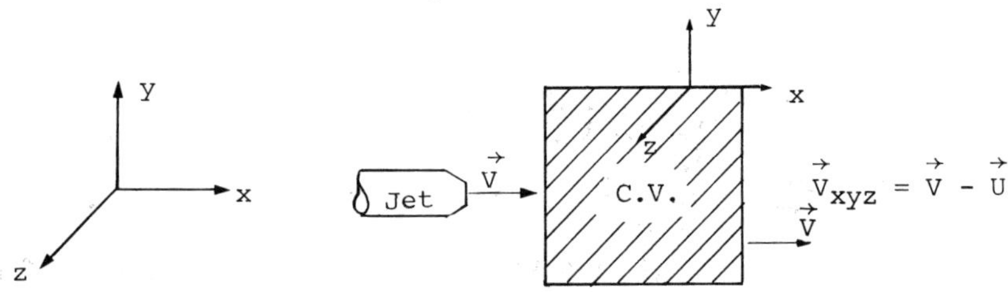

Moving control Volume with a constant velocity $\vec{U}$
Fig. 4.1

In the case of a steady one-dimensional flow shown in Figure 4.2, the equation is reduced to

$$F_x = \rho Q(V_{x2} - V_{x1}) \qquad (4.15)$$

The vector form of the equation (4.14) may be written as

$$F_i = \int_{CS} T_i dA + \int_{CV} B_i \rho d\forall$$

$$= \int_{CS} (\vec{V}_{xyz})_i \rho \vec{V}_{xyz} \, dA + \frac{\partial}{\partial t} \int_{CV} (\vec{V}_{xyz})_i \rho d\forall \quad (4.16)$$

for $\quad i = x, y, z$

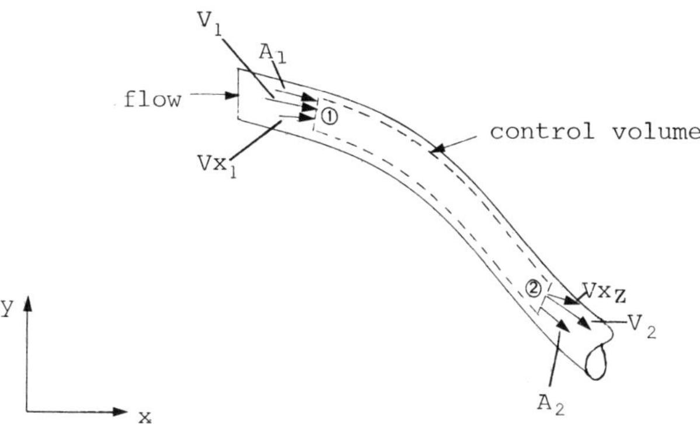

Fig 4.2 Control volume for a uniform flow normal to the control surface.

D) Linear Momentum Equation for Control Volume with Rectilinear Acceleration

This equation can be written as

$$\vec{F}_S + \vec{F}_B - \int_{CV} \vec{a}_{rf} \rho d\forall = \frac{\partial}{\partial t} \int_{CV} \vec{V}_{xyz} \rho d\forall$$

$$(4.17)$$

$$+ \int_{CS} \vec{V}_{xyz} \rho \vec{V}_{xyz} \, d\vec{A}$$

where

$$a_{rf} = \text{The acceleration of the control volume}$$
relative to a fixed reference coordinate xyz.

The vector form of the equation (4.17) may be written as

$$F_i = \int_{CS} T_i \, dA + \int_{CV} B_i \rho \, dV - \int_{CV} (\vec{a}_{rf})_i \rho \, dV$$

$$= \frac{\partial}{\partial t} \int_{CV} (v_{xyz})_i \rho \, dV + \int_{CS} (V_{xyz})_i \rho \vec{V}_{xyz} \cdot d\vec{A} \tag{4.18}$$

for $\qquad i = x, y, z$

Note: For an accelerating control volume one must label a coordinate system xyz on the control volume and a fixed reference coordinate xyz.

E) Linear Momentum Equation for Control Volume with Arbitrary Acceleration

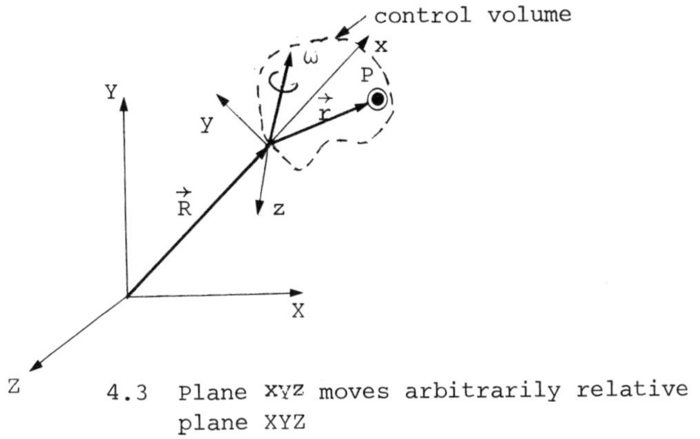

4.3　Plane xyz moves arbitrarily relative to plane XYZ

The most general control volume formulation of Newton's second law is

$$\vec{F}_S + \vec{F}_B - \int_{CV} [\vec{a}_{rf} + 2\omega \times \vec{V}_{xyz} + \omega \times (\vec{\omega} \times \vec{r}) + \dot{\vec{\omega}} \times \vec{r}] \rho \, dV$$

$$= \frac{\partial}{\partial t} \int_{CV} \vec{V}_{xyz} \rho \, dV + \int_{CS} \vec{V}_{xyz} \rho \vec{V}_{xyz} \, d\vec{A}$$

$$\tag{4.19}$$

where

$$\frac{\partial}{\partial t} \int \vec{V}_{xyz} \rho \, dV = \text{The sum of the rate of change of momentum in the control volume}$$

$$\int_{CS} \vec{V}_{xyz}\,\rho\,\vec{V}_{xyz}\cdot d\vec{A} = \text{The net rate of efflux of momentum}$$
through the control volume

$\vec{a}_{rf}$ = Absolute rectilinear acceleration of a moving reference xyz relative to a fixed coordinate (XYZ).

$2\,\vec{\omega}\times\vec{V}_{xyz}$ = Coriolis acceleration due to motion of the particle in the moving coordinate.

$\vec{\omega}\times(\vec{\omega}\times\vec{r})$ = Centrifugal acceleration due to rotation of the moving coordinate.

$\dot{\vec{\omega}}\times\vec{r}$ = Tangential acceleration due to angular acceleration of the moving reference.

F) Moment of Momentum Equation for Control Volume

a) Equation for Fixed Control Volume

The total torque that acts on the control volume is equal to the rate of efflux of angular momentum from the control volume plus the rate of change of angular momentum inside the control volume. The general vector equation is

$$\vec{r}\times\vec{F}_s + \int_{CV}\vec{r}\times\vec{g}\rho\,dV + \vec{T}_{shaft} = \int_{CS}\vec{r}\times\vec{V}\rho\vec{V}\cdot d\vec{A}$$

$$+ \frac{\partial}{\partial t}\int_{CV}\vec{r}\times\vec{V}\rho\,dV \qquad (4.20)$$

b) Equation for Rotating Control Volume

The general vector equation for the moment of momentum for a non-inertial control volume is

$$\vec{r}\times\vec{F}_s + \int_{CV}\vec{r}\times\vec{g}\rho\,dV + \vec{T}_{shaft} - \int_{CV}\vec{r}\times[2\vec{\omega}\times\vec{V}_{xyz}$$

$$+ \vec{\omega}\times(\vec{\omega}\times\vec{r}) + \dot{\vec{\omega}}\times\vec{r}]\rho\,dV$$

$$= \frac{\partial}{\partial t}\int_{CV}\vec{r}\times\vec{V}_{xyz}\rho\,dV + \int_{CS}\vec{r}\times\vec{V}_{xyz}\rho\vec{V}_{xyz}\,d\vec{A} \qquad (4.21)$$

G)   The First Law of Thermodynamics

The control volume formulation of the first law of thermodynamics is

$$\dot{Q} + \dot{W}_s + \dot{W}_{shear} + \dot{W}_{other}$$
$$= \frac{\partial}{\partial t} \int_{CV} e\rho\, d\forall + \int_{CS} \left(u + pv + \frac{v^2}{2} + gz\right)\rho\, \vec{v} \cdot d\vec{A} \tag{4.22}$$

where

$\dot{Q}$ = Heat transferred to the control volume

$\dot{W}_s$ = Rate of work crossing the control surface by shaft work

$\dot{W}_{shear}$ = Rate of work done by shear stresses

$\dot{W}_{other}$ = Any other work that can be added to the control volume (electrical energy, electromagnetic energy, etc.).

$\frac{\partial}{\partial t} \int_{CV} e\rho\, d\forall$ = The rate of change of energy in the control volume

$$e = \left(u + \frac{v^2}{2} + gz\right)$$

$\int_{CS} \left(u + pv + \frac{v^2}{2} + gz\right)\rho\, \vec{V} \cdot d\vec{A}$ = The net rate of energy through the control surface

H)   The Second Law of Thermodynamics

The control volume formulation of the second law of thermodynamics is

$$\frac{\partial}{\partial t} \int_{CV} s\rho\, d\forall + \int_{CS} s\rho\, \vec{V} \cdot d\vec{A} \geq \int_{CS} \frac{1}{T}\left(\frac{\dot{Q}}{A}\right) dA \tag{4.23}$$

CHAPTER 5

INTRODUCTION TO DIFFERENTIAL ANALYSIS OF FLUID MOTION

In Chapter 4 the basic equations in integral form for a control volume were developed. To obtain a detailed knowledge of the flow field the equations must also be developed in a differential form.

5.1 THE CONTINUITY EQUATION

A) In Rectangular Coordinates

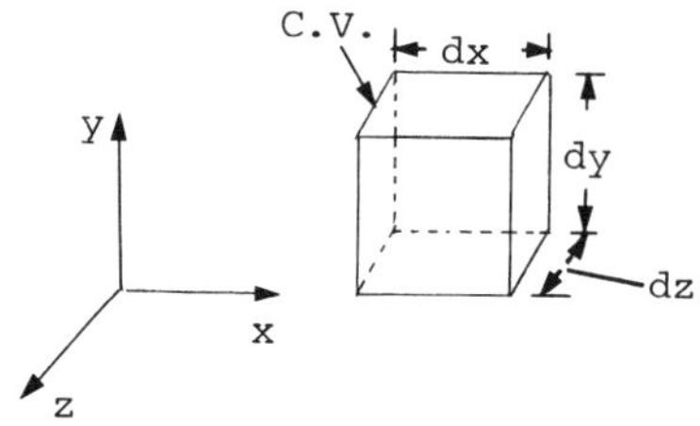

Fig. 5.1 Differential control volume
in rectangular coordinates

The differential form of the continuity equation is given by

$$\frac{\partial(\rho u)}{\partial x} + \frac{\partial(\rho v)}{\partial y} + \frac{\partial(\rho w)}{\partial z} = -\frac{\partial \rho}{\partial t} \qquad (5.1)$$

where u, v, and w are the velocity components in the x, y, and z directions, respectively. The continuity equation may be written more compactly in vector notation as

$$\nabla \cdot \rho \vec{v} = -\frac{\partial \rho}{\partial t}$$

(5.2)

where

$$\nabla = \hat{i} \frac{\partial}{\partial x} + \hat{j} \frac{\partial}{\partial y} + \hat{k} \frac{\partial}{\partial z}$$

For steady incompressible flow, ρ constant, and $\partial \rho / \partial t = 0$. Thus the continuity equation is reduced to

$$\frac{\partial u}{\partial x} + \frac{\partial v}{\partial y} + \frac{\partial w}{\partial z} = 0$$

(5.3)

or

$$\nabla \cdot \vec{v} = 0$$

(5.4)

and for steady compressible flow,

$$\frac{\partial(\rho u)}{\partial x} + \frac{\partial(\rho v)}{\partial y} + \frac{\partial(\rho w)}{\partial z} = 0$$

(5.5)

or,

$$\nabla \cdot \rho \vec{v} = 0$$

(5.6)

B) In a Cylindrical Coordinate System

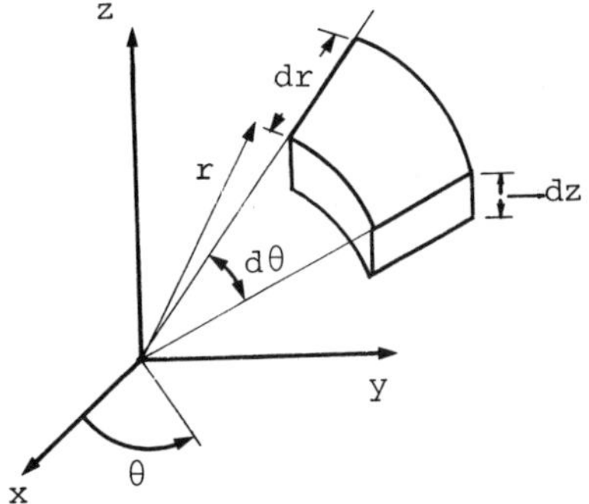

Fig. 5.2   Cylindrical coordinates

The differential form of the continuity equation in cylindrical coordinates is given by

$$\frac{1}{r} \frac{\partial r \rho V_r}{\partial r} + \frac{1}{r} \frac{\partial \rho V_\theta}{\partial \theta} + \frac{\partial \rho V_z}{\partial z} = -\frac{\partial \rho}{\partial t}$$

(5.7)

or,

$$\nabla \cdot \rho \vec{v} = - \frac{\partial \rho}{\partial t} \tag{5.8}$$

For steady incompressible flow, ρ = constant, and the continuity equation is reduced to

$$\frac{1}{r} \frac{\partial r V_r}{\partial r} + \frac{1}{r} \frac{\partial V}{\partial \theta} + \frac{\partial V_z}{\partial z} = 0 \tag{5.9}$$

and for steady compressible flow,

$$\frac{1}{r} \frac{\partial r \rho V_r}{\partial r} + \frac{1}{r} \frac{\partial \rho V_\theta}{\partial \theta} + \frac{\partial \rho V_z}{\partial z} = 0 \tag{5.10}$$

5.2 THE DIFFERENTIAL FORM OF MOMENTUM EQUATION

A) Acceleration of a Fluid Particle in a Velocity Field

For a velocity field $\vec{v} = \vec{v}(x,y,z,t)$, the acceleration $\left(\frac{D\vec{v}}{Dt}\right)$ of a fluid particle is given by

$$a = \frac{D\vec{v}}{Dt} = \frac{d}{dt} \vec{v}(x,y,z,t) = u \frac{\partial \vec{v}}{\partial x} + v \frac{\partial \vec{v}}{\partial y} + w \frac{\partial \vec{v}}{\partial z} + \frac{\partial \vec{v}}{\partial t} \tag{5.11}$$

where

$u \frac{\partial \vec{v}}{\partial x} + v \frac{\partial \vec{v}}{\partial y} + w \frac{\partial \vec{v}}{\partial z}$ = Time rate of change of velocity of the particle, due to its changing position in the field (acceleration of transport or convective acceleration)

u, v, w = Scalar velocity components of the particle in the x, y, and z directions respectively

$\frac{\partial \vec{v}}{\partial t}$ = Rate of change of the velocity field itself at the position occupied by the particle at time t (local acceleration)

B) Forces Acting on a Fluid Particle

Forces acting on a fluid element may be classified as

body forces and surface forces. Surface forces include both normal forces and tangential (shear) forces (Fig. 5.3).

The net force in the x-direction is given by:

$$dF_x = dF_{sx} + dF_{Bx} = \left(\rho B_x + \frac{\partial \sigma_{xx}}{\partial x} + \frac{\partial \tau_{yx}}{\partial y} + \frac{\partial \tau_{zx}}{\partial z} \right) dx\,dy\,dz \qquad (5.12)$$

In the y-direction

$$dF_y = dF_{sy} + dF_{BY} = \left(\rho B_y + \frac{\partial \tau_{xy}}{\partial x} + \frac{\partial \sigma_{yy}}{\partial y} + \frac{\partial \tau_{zy}}{\partial z} \right) dx\,dy\,dz \qquad (5.13)$$

In the z-direction

$$dF_z = dF_{sz} + dF_{Bz} = \left(\rho B_z + \frac{\partial \tau_{xz}}{\partial x} + \frac{\partial \tau_{yz}}{\partial y} + \frac{\partial \sigma_{zz}}{\partial z} \right) dx\,dy\,dz \qquad (5.14)$$

where B is the body force per unit of mass.

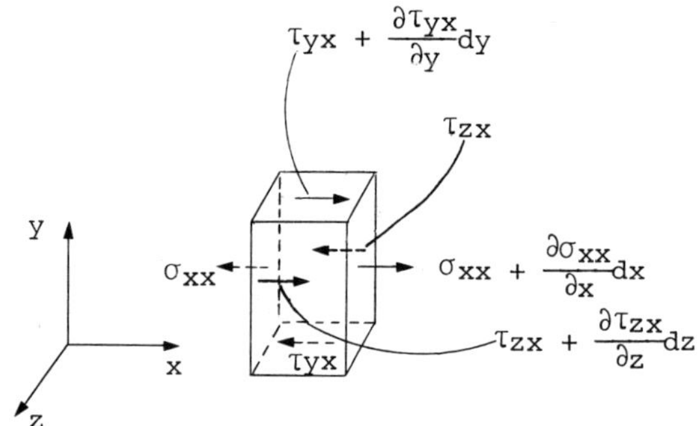

Fig. 5.3   Stresses in the x direction

C) Differential Momentum Equation

For a system of mass, dm, Newton's second law can be written as

$$d\vec{F} = dm \, \frac{d\vec{v}}{dt} \qquad (5.15)$$

By substituting the expressions of acceleration $\left(\frac{d\vec{V}}{dt} \right)$ and the expressions of force $(d\vec{F})$ acting on the element of mass (dm) in the above equation, we can obtain the differential equations of motion which are:

$$(5.16)$$

$$\rho B_x + \frac{\partial \sigma_{xx}}{\partial x} + \frac{\partial \tau_{yx}}{\partial y} + \frac{\partial \tau_{zx}}{\partial z} = \rho\left(\frac{\partial u}{\partial t} + u\frac{\partial u}{\partial x} + \sigma\frac{\partial u}{\partial y} + w\frac{\partial u}{\partial z}\right)$$

$$(5.17)$$

$$\rho B_y + \frac{\partial \tau_{xy}}{\partial x} + \frac{\partial \sigma_{yy}}{\partial y} + \frac{\partial \tau_{zy}}{\partial z} = \rho\left(\frac{\partial v}{\partial t} + u\frac{\partial v}{\partial x} + \sigma\frac{\partial \sigma}{\partial y} + w\frac{\partial \sigma}{\partial z}\right)$$

$$(5.18)$$

$$\rho B_z + \frac{\partial \tau_{xz}}{\partial x} + \frac{\partial \tau_{yz}}{\partial y} + \frac{\partial \sigma_{zz}}{\partial z} = \rho\left(\frac{\partial w}{\partial t} + u\frac{\partial w}{\partial x} + v\frac{\partial w}{\partial y} + w\frac{\partial w}{\partial z}\right)$$

For Newtonian fluids the stresses may be expressed in terms of velocity gradients and fluid properties as

$$\tau_{xy} = \tau_{yx} = \mu\left(\frac{\partial \sigma}{\partial x} + \frac{\partial u}{\partial y}\right) \qquad (5.19)$$

$$\tau_{yz} = \tau_{zy} = \mu\left(\frac{\partial w}{\partial y} + \frac{\partial v}{\partial z}\right) \qquad (5.20)$$

$$\tau_{zx} = \tau_{xz} = \mu\left(\frac{\partial u}{\partial z} + \frac{\partial w}{\partial x}\right) \qquad (5.21)$$

$$\sigma_{xx} = -p - \frac{2}{3}\mu\nabla\cdot\vec{V} + 2\mu\frac{\partial u}{\partial x} \qquad (5.22)$$

$$\sigma_{xy} = -p - \frac{2}{3}\mu\nabla\cdot\vec{V} + 2\mu\frac{\partial v}{\partial y} \qquad (5.23)$$

$$\sigma_{zz} = -p - \frac{2}{3}\mu\nabla\cdot\vec{V} + 2\mu\frac{\partial w}{\partial z} \qquad (5.24)$$

where p is the local pressure.

If these equations are introduced into the differential equations of motion, we can obtain the Navier-Stokes equations, which are:

$$\rho\left(\frac{\partial u}{\partial t} + u\frac{\partial u}{\partial x} + v\frac{\partial u}{\partial y} + w\frac{\partial u}{\partial z}\right) = \rho B_x - \frac{\partial p}{\partial x} + \mu\left(\frac{\partial^2 u}{\partial x^2} + \frac{\partial^2 u}{\partial y^2} + \frac{\partial^2 u}{\partial z^2}\right)$$

$$(5.25)$$

$$\rho\left(\frac{\partial v}{\partial t} + u\frac{\partial v}{\partial x} + v\frac{\partial v}{\partial y} + w\frac{\partial v}{\partial z}\right) = \rho B_y - \frac{\partial p}{\partial y} + \mu\left(\frac{\partial^2 v}{\partial x^2} + \frac{\partial^2 v}{\partial y^2} + \frac{\partial^2 v}{\partial z^2}\right)$$

$$(5.26)$$

$$\rho \left(\frac{\partial w}{\partial t} + u \frac{\partial w}{\partial x} + v \frac{\partial w}{\partial y} + w \frac{\partial w}{\partial z} \right) = \rho B_z - \frac{\partial p}{\partial z} + \mu \left(\frac{\partial^2 w}{\partial x^2} + \frac{\partial^2 w}{\partial y^2} + \frac{\partial^2 w}{\partial z^2} \right)$$

$$(5.27)$$

5.3 EULER'S EQUATIONS

Euler's equations are obtained from the equations of motion asuming a frictionless flow. Since, in a frictionless flow, there can be no shear stress present, the surface forces are due to pressure.

Thus, the Euler's equations are

$$\rho B_x - \frac{\partial p}{\partial x} = \rho \left(\frac{\partial u}{\partial t} + u \frac{\partial u}{\partial x} + v \frac{\partial u}{\partial y} + w \frac{\partial u}{\partial z} \right) \qquad (5.28)$$

$$\rho B_y - \frac{\partial p}{\partial y} = \rho \left(\frac{\partial v}{\partial t} + u \frac{\partial v}{\partial x} + v \frac{\partial v}{\partial y} + w \frac{\partial v}{\partial z} \right) \qquad (5.29)$$

$$\rho B_z - \frac{\partial p}{\partial z} = \rho \left(\frac{\partial w}{\partial t} + u \frac{\partial w}{\partial x} + v \frac{\partial w}{\partial y} + w \frac{\partial w}{\partial z} \right) \qquad (5.30)$$

These equations can also be expressed in a vectorial form as

$$\rho \vec{B} - \nabla p = \rho \left(\frac{\partial \vec{V}}{\partial t} + u \frac{\partial \vec{V}}{\partial x} + v \frac{\partial \vec{V}}{\partial y} + w \frac{\partial \vec{V}}{\partial z} \right)$$

or,

$$\rho \vec{B} - \nabla p = \rho \frac{D\vec{V}}{Dt} \qquad (5.31)$$

If the only body force is due to gravity, Euler's equation becomes

$$\rho \vec{g} - \nabla p = \rho \frac{D\vec{V}}{Dt} \qquad (5.32)$$

For frictionless flow along a streamline the Euler's equation is reduced to

$$\frac{dp}{\rho} + gdz + vdv = 0$$

In cylindrical coordinates, Euler's equation (with gravity the only force) becomes

$$g_r - \frac{1}{\rho}\frac{\partial p}{\partial r} = a_r = \frac{\partial v_r}{\partial t} + v_r\frac{\partial v_r}{\partial r} + \frac{v_\theta}{r}\frac{\partial v_r}{\partial \theta} + v_z\frac{\partial v_r}{\partial z} - \frac{v_\theta^2}{r} \qquad (5.33)$$

$$g_\theta - \frac{1}{\rho r}\frac{\partial p}{\partial \theta} = a_\theta = \frac{\partial v_\theta}{\partial t} + v_r\frac{\partial v_\theta}{\partial r} + \frac{v_\theta}{r}\frac{\partial v_\theta}{\partial \theta} + v_z\frac{\partial v_\theta}{\partial z} + \frac{v_r v_\theta}{r} \qquad (5.34)$$

$$g_z - \frac{1}{\rho}\frac{\partial p}{\partial z} = a_z = \frac{\partial v_z}{\partial t} + v_r\frac{\partial v_z}{\partial r} + \frac{v_\theta}{r}\frac{\partial v_z}{\partial \theta} + v_z\frac{\partial v_z}{\partial z} \qquad (5.35)$$

5.4 BERNOULLI EQUATION

By integrating Euler's equation we get the Bernoulli equation:

$$\frac{p}{\rho} + gz + \frac{v^2}{2} = \text{constant} \qquad (5.36)$$

(along a streamline)

The assumptions by which the Bernoulli equation holds, are:

A) Steady flow

B) Incompressible flow

C) Frictionless flow

D) Flow along a streamline

Applied between any two points on a streamline the Bernoulli equation becomes

$$\frac{p_1}{\rho} + \frac{v_1^2}{2} + gz_1 = \frac{p_2}{\rho} + \frac{v_2^2}{2} + gz_2$$

or

$$\frac{p_1}{\gamma} + \frac{v_1^2}{2g} + z_1 = \frac{p_2}{\gamma} + \frac{v_2^2}{2g} + z_2 \qquad (5.37)$$

The Bernoulli equation can also be derived by using the first law of thermodynamics.

The first law of thermodynamics is reduced to Bernoulli's equation under the following restrictions:

A) $\dot{W}_s = \dot{W}_{shear} = \dot{W}_{other} = 0$

B) Steady flow

C) Uniform flow at each section

D) Incompressible flow

E) $u_2 - u_1 = \dot{Q}/\dot{m}$

CHAPTER 6

CONSIDERATIONS FOR COMPRESSIBLE FLOW

6.1 REVIEW FOR THERMODYNAMICS

A) Equation of State

The pressure, density, and temperature of a substance may be related by the equation of state:

$$p = \rho R T \qquad (6.1)$$

where

$$R = \frac{R_u}{M}$$

R_u = Universal gas constant

$\quad$ = 8314 Nm /kg mole K

M = Molecular mass of gas

Substances that satisfy this equation are known as perfect gases.

B) Internal Energy

For any substance that follows the equation of state $p = \rho R T$, the internal energy, 'u', is given by the equation $du = c_v dT$, or by integrating this equation with constant specific heat

41

$$u_2 - u_1 = \int_{u_1}^{u_2} du = \int_{T_1}^{T_2} c_v \, dT = c_v(T_2 - T_1) \qquad (6.2)$$

C) Enthalpy

The enthalpy of a substance is defined as $h = u + pv$, and for an ideal gas, $dh = c_p \, dT$, and when integrating with constant specific heat,

$$h_2 - h_1 = \int_{h_1}^{h_2} dh = \int_{T_1}^{T_2} c_p \, dT = c_p(T_2 - T_1) \qquad (6.3)$$

D) Entropy

Entropy is defined by the equation

$$\Delta s = \int_{rev} \frac{\delta Q}{T} \quad \text{or,} \quad \Delta s = \left(\frac{\delta Q}{T}\right)_{rev} \qquad (6.4)$$

The inequality of Clausius states that

$$\oint \frac{\delta Q}{T} \leq 0 \qquad (6.5)$$

As a consequence of the second law the results can be extended to

$$Tds \geq \delta Q \qquad (6.6)$$

For reversible processes,

$$Tds = \frac{\delta Q}{dm} \qquad (6.7)$$

For irreversible processes,

$$Tds > \frac{\delta Q}{dm} \qquad (6.8)$$

For an adiabatic process, $\delta Q/\delta m = 0$. Thus,

$$ds = 0 \quad \text{(reversible adiabatic process)}$$

$$ds > 0 \quad \text{(irreversible adiabatic process)}$$

For an ideal gas with constant specific heats, we have

$$\boxed{\begin{aligned} s_2 - s_1 \;&=\; c_v \ln \frac{T_2}{T_1} + R\ln \frac{V_2}{V_1} \\[2ex] &=\; c_p \ln \frac{T_2}{T_1} - R\ln \frac{p_2}{p_1} \end{aligned}}$$

$$(6.9)$$

$$(6.10)$$

E) Specific Heats

The constant pressure of specific heat, c_p, is defined as

$$c_p = \left(\frac{\partial h}{\partial T}\right)_p \tag{6.11}$$

The constant volume of specific heat, c_v, is defined as

$$c_v = \left(\frac{\partial u}{\partial T}\right)_v \tag{6.12}$$

Also, for an ideal gas we have the following relations:

$$c_p - c_v = R \tag{6.13}$$

$$K = \frac{c_p}{c_v} \tag{6.14}$$

$$c_p = \frac{KR}{K-1} \tag{6.15}$$

$$c_v = \frac{R}{K-1} \tag{6.16}$$

6.2 PROPAGATION OF SOUND WAVES

A) Speed of Sound; the Mach Number

The Mach number is defined as:

$$M = \frac{V}{c} \tag{6.17}$$

where V = Local flow speed

           c = Local speed of sound

For an ideal gas the speed of sound is given by the relation

$$c = \sqrt{KRT} \qquad (6.18)$$

where

$$R = \text{A constant for each gas}$$

$$T = \text{Temperature of the gas}$$

$$K = \frac{c_p}{c_v}$$

B) Types of Flow; The Mach Cone

The types of flow are shown in the following table

$M < 1$	subsonic
$M = 1$	sonic
$M > 1$	supersonic
$M \overset{\sim}{>} 5$	hypersonic

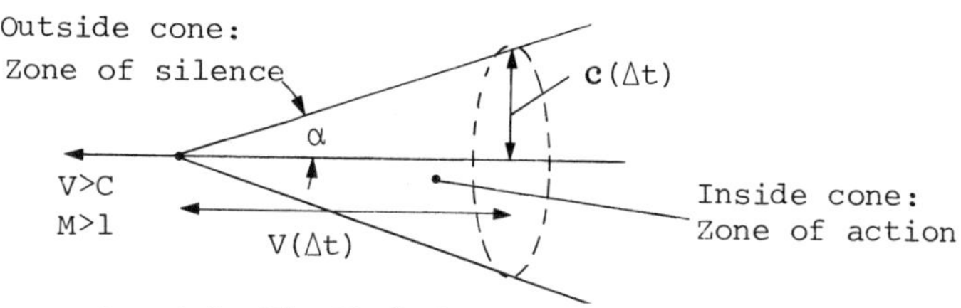

Fig. 6.1   The Mach Cone

The cone angle, α , can be related to the Mach number at which the source moves. From the geometry of (Fig. 6.1), we conclude that

$$\sin \alpha = \frac{c}{V} = \frac{1}{M} \qquad \text{or,}$$

$$\alpha = \sin^{-1}\left(\frac{1}{M}\right) \qquad (6.19)$$

where α is the half-angle of the Mach cone.

The regions inside and outside the cone are sometimes called the zone of action and the zone of silence, respectively.

6.3 LOCAL ISENTROPIC STAGNATION PROPERTIES

Local isentropic stagnation properties are properties that would be obtained at any point in a flow field if the fluid at that point were decelerated from local conditions to zero velocity following an isentropic process.

The equations for the isentropic stagnation properties of an ideal gas are:

$$\frac{p_0}{p} = \left[1 + \frac{K-1}{2} M^2 \right]^{K/(K-1)} \tag{6.20}$$

$$\frac{T_0}{T} = 1 + \frac{K-1}{2} M^2$$

$$\frac{\rho_0}{\rho} = \left[1 + \frac{K-1}{2} M^2 \right]^{1/(K-1)} \tag{6.21}$$

For ideal gases, the ratios of local isentropic stagnation properties to corresponding static properties for an ideal gas can be obtained from standard textbooks.

6.4 CRITICAL CONDITION

Critical conditions are those conditions where the Mach number is equal to unity (sonic conditions).

Sonic conditions are marked with an asterisk:

$$V^* \equiv C^*$$

At critical conditions the stagnation properties become

$$\frac{p_0^*}{p^*} = \left[1 + \frac{K-1}{2} \right]^{K/(K-1)} \tag{6.22}$$

$$\frac{T_0^*}{T^*} = 1 + \frac{K-1}{2} \qquad (6.23)$$

$$\frac{\rho_0^*}{\rho^*} = \left[1 + \frac{K-1}{2}\right]^{1/(K-1)} \qquad (6.24)$$

The critical speed may be written in terms of either the critical temperature, T^*, or the critical stagnation temperature, T_0^*.

$$V^* = c^* = \sqrt{\frac{2K}{K+1}\, R T_0^*} \qquad (6.25)$$

$$V^* = c^* = \sqrt{K R T^*} \qquad (6.26)$$

CHAPTER 7

ONE-DIMENSIONAL COMPRESSIBLE FLOW

7.1 BASIC EQUATIONS FOR ISENTROPIC FLOW

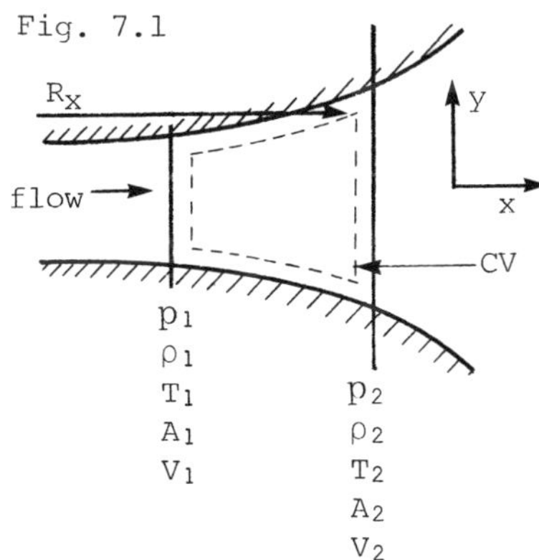

The basic equations applied to the control volume shown in Figure 7.1 are

A) Continuity Equation

$$\frac{\partial}{\partial t} \int_{cv} \rho \, d\forall + \int_{cs} \rho \, \vec{V} \cdot d\vec{A} = 0$$

or

$$-\rho_1 V_1 A_1 + \rho_2 V_2 A_2 = 0$$

or

$$\boxed{\rho_1 V_1 A_1 = \rho_2 V_2 A_2 = \rho V A = \dot{m} = \text{constant}} \qquad (7.1)$$

B) Momentum Equation

$$\vec{F}_{S_x} + \vec{F}_{B_x} = \int_{cs} \vec{V} \rho \vec{V} dA + \frac{\partial}{\partial t} \int_{cv} \overset{0}{\cancel{\vec{V} \rho d\rlap{--}V}}$$

or

$$R_x + p_1 A_1 - p_2 A_2 = -V_1^2 \rho_1 A_1 + V_2^2 \rho_2 A_2$$

or

$$R_x + p_1 A_1 - p_2 A_2 = \dot{m} V_2 - \dot{m} V_1 \qquad (7.2)$$

where

$$R_x = \text{External force acting on the control volume}$$

C) First Law of Thermodynamics

$$\overset{0}{\cancel{\dot{Q}}} + \overset{0}{\cancel{\dot{W}_s}} + \overset{0}{\cancel{\dot{W}_{shear}}} + \dot{W}_{other} = \frac{\partial}{\partial t} \int_{cv} \overset{0}{\cancel{e \rho \, d\rlap{--}V}}$$

$$+ \int_{cs} (e + pv) \rho \, \vec{v} \cdot d\vec{A}$$

where

$$e = u + \frac{V^2}{2} + \overset{0}{\cancel{gz}}$$

Then the first law of thermodynamics becomes

$$\boxed{h_1 + \frac{V_1^2}{2} = h_2 + \frac{V_2^2}{2} = \text{constant}} \qquad (7.3)$$

where

$$h = u + pv$$

D) Second Law of Thermodynamics

$$\int_{cs} \frac{1}{T} \frac{\dot{Q}}{A} dA \leq \frac{\partial}{\partial t} \int_{cv} S \rho \, d\rlap{--}V + \int_{cs} S \rho \vec{V} \cdot d\vec{A}$$

or

$$s_1[-\rho_1 V_1 A_1] + s_2[\rho_2 V_2 A_2] = 0$$

or

$$\boxed{s_1 = s_2 = s = \text{constant}}$$
(7.4)

E) Equation of State

$$p = \rho RT$$
(7.5)

7.2 EFFECTS OF AREA VARIATION ON FLOW PROPERTIES IN ISENTROPIC FLOW

In order to study the effects of area variation on flow properties in isentropic flow, the following equations are considered:

$$\boxed{\dfrac{dA}{A} = \dfrac{dP}{\rho V^2}(1-M^2)}$$
(7.6a)

$$\boxed{\dfrac{dA}{A} = \dfrac{-dV}{V}(1-M^2)}$$
(7.6b)

From these equations, any area variation will result in a positive or a negative change of velocity and pressure depending upon the M value:

for $M^2 < 1$, $dP > 0$ and $dV < 0$

for $M^2 > 1$, $dP < 0$ and $dV > 0$

The results are summarized in Figure 7.2.

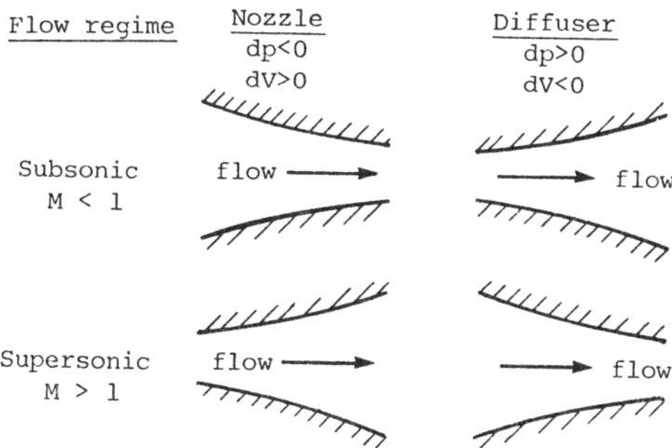

Fig. 7.2 Nozzle and diffuser shapes as a function of initial Mach number.

7.3 ISENTROPIC FLOW OF AN IDEAL GAS

A) Basic Equations

Continuity:	$\rho_1 V_1 A_1 = \rho_2 V_2 A_2 = \dot{m}$	(7.7)
Momentum:	$R_x + p_1 A_1 - p_2 A_2 = \dot{m} V_2 - \dot{m} V_1$	(7.8)
First Law:	$h_1 + \dfrac{V_1^2}{2} = h_2 + \dfrac{V_2^2}{2} = h + \dfrac{V^2}{2}$	(7.9)
Second Law:	$s_1 = s_2 = S$	(7.10)
Equation of State:	$p = \rho R T$	(7.11)
Process Equation:	$p/\rho^k = \text{constant}$	(7.12)

B) Stagnation Properties and Critical Properties for Isentropic Flow of an Ideal Gas

Stagnation pressure:	$\dfrac{p_0}{p} = \left[1 + \dfrac{K-1}{2} M^2\right]^{K/(K-1)}$	(7.13)
Stagnation temperature:	$\dfrac{T_0}{T} = 1 + \dfrac{K-1}{2} M^2$	(7.14)
Stagnation density:	$\dfrac{\rho_0}{\rho} = \left[1 + \dfrac{K-1}{2} M^2\right]^{1/(K-1)}$	(7.15)

At critical conditions the stagnation properties become (K = 1.4)

$$\frac{p_0}{p^*} = \left[1 + \frac{K-1}{2}\right]^{K/(K-1)} = 1.893 \qquad (7.16)$$

$$\frac{T_0}{T^*} = 1 + \frac{K-1}{2} = 1.200 \qquad (7.17)$$

$$\frac{\rho_0}{\rho^*} = \left[1 + \frac{K-1}{2}\right]^{1/(K-1)} = 1.577 \qquad (7.18)$$

In addition,

$$V^* = c^* = \sqrt{\frac{2K}{K+1}\, R T_0}$$

(7.19)

$$\frac{A}{A^*} = \frac{1}{M}\left[\frac{1 + \frac{K-1}{2}M^2}{1 + \frac{K-1}{2}}\right]^{(K+1)/2(K-1)}$$

(7.20)

since, $\rho\, AV = \text{constant}$

7.4 ISENTROPIC FLOW IN A CONVERGING NOZZLE

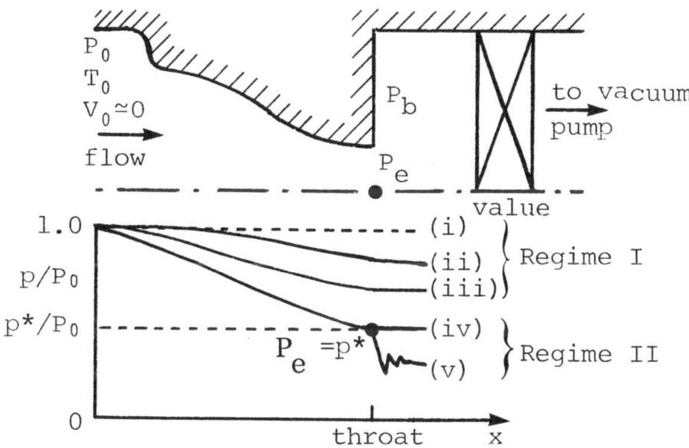

Fig. 7.3 Converging nozzle operating at various
      back pressures.

The effects of variations in back pressure, p_b, on the pressure distribution through a converging nozzle are graphically illustrated in Figure 7.3. The flow through a converging nozzle may be divided into two regimes:

REGIME I :

In this regime when the back pressure, P_b, decreases, the flowrate increases and the exit plane pressure, P_e, decreases (see curves b and c, Fig. 7.3). The flow at the exit plane will eventually reach a Mach number equal to unity (curve d) where the pressure is the critical pressure, $P*$ ($M = 1$, $P_b/P_0 = P*/P_0$).

REGIME II :

In this regime, a reduction of the back pressure, P_b, below the critical pressure, $P*$, has no effect on the flow conditions inside the nozzle. If $P_b \leq P*$, the nozzle is said to be choked. If $P_b \leq P*$, the flow leaving the nozzle expands to reach a lower back pressure (curve e). Figure 7.3 shows the T–S diagram for the flow processes in this regime.

7.5 ADIABATIC FLOW IN A CONSTANT AREA DUCT WITH FRICTION

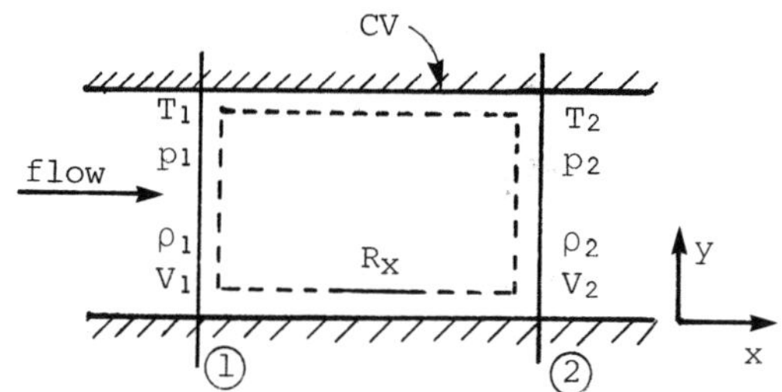

Fig. 7.4 Control volume used for integral analysis of frictional adiabatic flow.

The basic equations for the steady flow of an ideal gas, with constant specific heats applied to the control volume shown in Figure 7.4 are:

A) Continuity Equation

$$0 = \frac{\partial}{\partial t} \int_{cv} \rho \, dV + \int_{cs} \rho \vec{V} d\vec{A}$$

Because of a steady flow at each section, the continuity continuity equation becomes

$$\rho_1 V_1 = \rho_2 V_2 = \text{constant} \tag{7.21}$$

B) Momentum Equation

$$F_{sx} + F_{Bx} = \frac{\partial}{\partial t} \int_{cv} \cancel{V_x \rho d\mathcal{V}}^{0} + \int_{cs} V_x \rho \vec{V} d\vec{A}$$

if $\quad F_{Bx} = 0$ and $A_2 = A_1 = A$ then,

$$R_x + p_1 A - p_2 A = \dot{m} V_2 - \dot{m} V_1 \tag{7.22}$$

C) First Law of Thermodynamics

$$\cancel{\dot{Q}}^{0} + \cancel{\dot{W}_s}^{0} + \cancel{\dot{W}_{shear}}^{0} + \cancel{\dot{W}_{other}}^{0}$$

$$= \cancel{\frac{\partial}{\partial t} \int_{cv} e \rho d\mathcal{V}}^{0} + \int_{cs} (e + pv) \rho \vec{V} d\vec{A}$$

where

$$e = u + \frac{V^2}{2} + \cancel{g z}^{0}$$

or

$$h_2 - h_1 + \frac{V_2^2 - V_1^2}{2} = 0$$

where

$$h_2 - h_1 = C_p(T_2 - T_1)$$

or

$$h_1 + \frac{V_1^2}{2} = h_2 + \frac{V_2^2}{2} = \text{constant}$$

or

$$h_{01} = h_{02} \tag{7.23}$$

D) Second Law of Thermodynamics

$$\int_{cs} \frac{1}{T} \cancel{\frac{\dot{Q}}{A}}^{0} dA \leq \frac{\partial}{\partial t} \int_{cv} s \rho d\mathcal{V} + \int_{cs} s \rho \vec{V} d\vec{A}$$

Because the flow is frictional and irreversible,

$$\dot{m}(s_2 - s_1) > 0 \tag{7.24}$$

where

$$s_2 - s_1 = C_p \ln \frac{T_2}{T_1} - R \ln \frac{p_2}{p_1}$$

E) Equation of State

$$p = \rho RT \qquad (7.25)$$

7.6 THE FANNO LINE

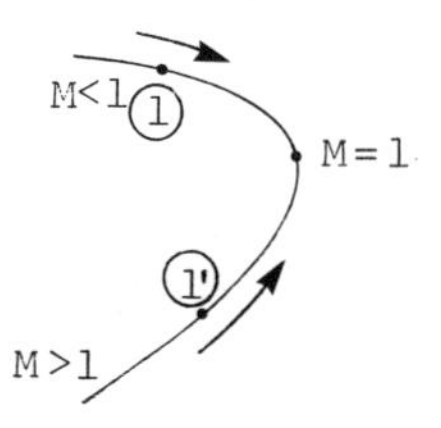

Fig. 7.5

Let us consider the governing equations developed in the last section and the relationship between h and T for an ideal gas (C_p = constant):

$$\Delta h = C_p(T_2 - T_1) \qquad (7.26)$$

Thus, we have six equations and seven unknowns. If the initial state is known, then there is an infinite number of final states (six equations and seven unknowns). The locus of the possible final states plotted on the T-S or h-S diagram is called the Fanno Line (Fig. 7.5). The Mach number is less than 1 (M<1) on the upper portion of the curve and it increases as we move to the right. The Mach number is greater than 1 (M>1) on the lower portion of the curve and it decreases as we move to the right. The Mach number is 1 (M=1) at the point of maximum entropy.

The effects of friction on flow properties in Fanno line flow are summarized in the following table:

Table 7.1

Property	M < 1	M > 1
Stagnation Temperature, T_0	Constant	Constant
Entropy, s	Increases	Increases
Stagnation Pressure, P_0	Decreases	Decreases
Temperature, T	Decreases	Increases
Velocity, V	Increases	Decreases
Mach number, M	Increases	Decreases
Density, ρ	Decreases	Increases
Pressure, p	Decreases	Increases
Stagnation Enthalpy	Constant	Constant

7.7 FRICTIONLESS FLOW IN A CONSTANT AREA DUCT WITH HEAT ADDITION

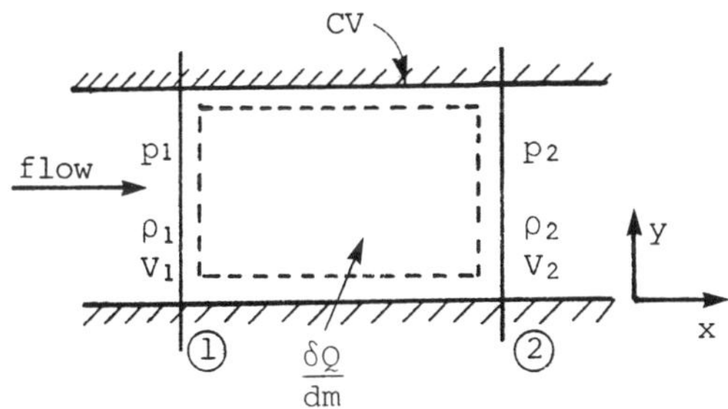

Fig. 7.6 Control volume used for integral analysis analysis of frictionless flow with heat transfer.

The basic equations for the steady, one-dimensional, frictionless flow of an ideal gas with constant specific heats applied to the control volume shown in (Fig. 7.6) are:

A) Continuity Equation

$$0 = \frac{\partial}{\partial t} \int_{cv} \rho d V + \int_{cs} \rho \vec{V} d\vec{A}$$

Because of a steady uniform flow at each section, the continuity equation becomes

$$\rho_1 V_1 \; = \; \rho_2 V_2 \; = \frac{\dot{m}}{A} \qquad\qquad (7.27)$$

B) Momentum Equation

$$F_{Sx} + \cancel{F_{Bx}}^{\,0} = \frac{\partial}{\partial t} \cancel{\int_{cv} V_x \,\rho\, d\mathcal{V}}^{\,0} + \int_{cs} V_x \,\rho\, \vec{V} d\vec{A} \quad \text{or,}$$

$$p_1 + \rho_1 v_1^{\,2} = p_2 + \rho_2 v_2^{\,2}$$

or

$$p + \rho\, v^2 = \text{constant} \qquad\qquad (7.28)$$

C) First Law of Thermodynamics

$$\dot{Q} + \cancel{\dot{W}_s}^{\,0} + \cancel{\dot{W}_{shear}}^{\,0} + \cancel{\dot{W}_{other}}^{\,0} = \frac{\partial}{\partial t}\cancel{\int_{cv} e\rho\, d\mathcal{V}}^{\,0} + \int_{cs} (e+pv)\rho\,\vec{v}d\vec{A}$$

where

$$e = u + \frac{V^2}{2} + \cancel{gz}^{\,0}$$

or

$$\frac{\delta Q}{dm} = \left[h_2 + \frac{V_2^{\,2}}{2} \right] - \left[h_1 + \frac{V_1^{\,2}}{2} \right] = (h_2 - h_1) + \left(\frac{V_2^{\,2} - V_1^{\,2}}{2} \right)$$

where

$$h_2 - h_1 = C_p(T_2 - T_1)$$

or

$$\boxed{\frac{\delta Q}{dm} = h_{02} - h_{01}} \qquad\qquad (7.29)$$

where

$$h_{01} = h_1 + \frac{V_1^{\,2}}{2}$$

$$h_{02} = h_2 + \frac{V_2^{\,2}}{2}$$

D) Second Law of Thermodynamics

$$\int_{cs} \frac{1}{T} \frac{\dot{Q}}{A} \, dA \leq \frac{\partial}{\partial t} \int_{cv} \overset{0}{\cancel{S\rho \, dV}} + \int_{cs} s\rho \vec{V}d\vec{A}$$

or

$$\int_{cs} \frac{1}{T} \frac{\dot{Q}}{A} \, dA \leq \dot{m}(s_2 - s_1) \qquad (7.30)$$

where

$$\boxed{s_2 - s_1 = C_p \ln \frac{T_2}{T_1} - R\ln \frac{p_2}{p_1}}$$

E) Equation of State

$$p = \rho RT \qquad (7.31)$$

7.8 THE RAYLEIGH LINE

Let us consider the governing equations (7.27) to (7.31) developed in the last section and equation (7.26); thus, we have six equations and seven unknowns. If the initial state is known then there are an infinite number of final states. The locus of the possible final states is plotted on the T-S or h-S diagram is called the Rayleigh line (Fig. 7-7). The mach number is less than 1 (M<1) on the upper portion of the curve and it increases as we move to the right. The Mach number is greater than one (M>1) on the lower portion of the curve and it decreases as we move to the right. The Mach number is one (M=1) at the point of maximum entropy (point b).

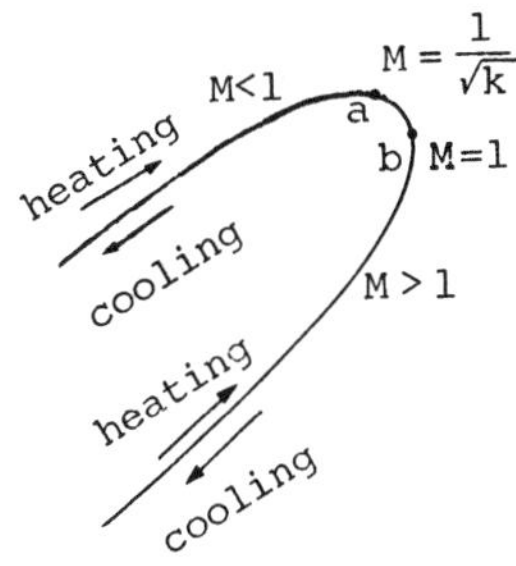

Fig. 7.7 Schematic Ts diagram for frictionless flow in a constant area duct with heat transfer (Rayleigh line flow).

The effects of heat transfer on properties in the steady, frictionless compressible flow of an ideal gas are summarized in the following table.

Table 7.2

	Heating		Cooling	
Property	M < 1	M > 1	M < 1	M > 1
Entropy, S	Increases	Increases	Decreases	Decreases
Stagnation Temperature, T_0	Increases	Increases	Decreases	Decreases
Temperature, T	$M < 1/\sqrt{K}$			
	Increases	Increases	Decreases	Decreases
	$1/\sqrt{K} < M < 1$			
	Decreases	Increases	Increases	Decreases
Mach number, M	Increases	Decreases	Decreases	Increases
Pressure, p	Decreases	Increases	Increases	Decreases
Velocity, V	Increases	Decreases	Decreases	Increases
Density, ρ	Decreases	Increases	Increases	Decreases
Stagnation Pressure, p_0	Decreases	Decreases	Increases	Increases

7.9 NORMAL SHOCK

Normal shocks are discontinuities of flow, which are normal to the flow direction and can occur in any supersonic flow field. The basic equations applied to the control volume around a normal shock (Fig. 7.8) are:

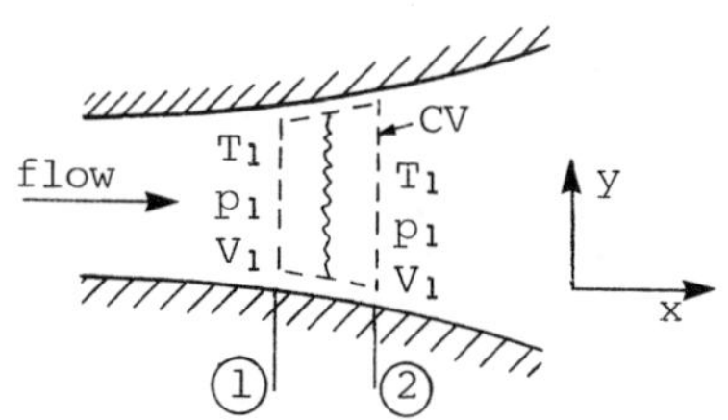

Fig. 7.8 Control volume used for analysis of normal shock.

A) Continuity Equation

$$0 = \frac{\partial}{\partial t} \int_{cv} \rho \, \cancel{dV}^{\,0} + \int_{cs} \rho \vec{V} d\vec{A}$$

Because of a steady uniform flow and, assuming that the area of change across the shock wave is negligible, (the shock is very thin, 0.2 microns), the continuity equation becomes

$$\rho_1 V_1 = \rho_2 V_2 = G \tag{7.32}$$

B) Momentum Equation

$$F_{Sx} + \cancel{F_{Bx}}^{\,0} = \cancel{\frac{\partial}{\partial t} \int_{cv} V_x p \, dV}^{\,0} + \int_{cs} V_x \rho \vec{V} d\vec{A}$$

on

$$F_{Sx} = \dot{m}V_2 - \dot{m}V_1$$

where

$$F_{Sx} = p_1 A - p_2 A$$

Finally, the momentum equation becomes

$$\boxed{p_1 + \rho_1 V_1^2 = p_2 + \rho_2 V_2^2} \tag{7.33}$$

C) First Law of Thermodynamics

$$\cancel{\dot{Q}}^{\,0} + \cancel{\dot{W}_s}^{\,0} + \cancel{\dot{W}_{shear}}^{\,0} + \dot{W}_{other}$$

$$= \frac{\partial}{\partial t} \int_{cv} e \rho \, dV + \int_{cs} (e+pv) \rho \vec{v} d\vec{A}$$

where

$$e = u + \frac{V^2}{2} + \cancel{gz}^{\,0}$$

Finally, the first law becomes

$$h_1 + \frac{V_1^2}{2} = h_2 + \frac{V_2^2}{2} \tag{7.34a}$$

or, in terms of stagnation enthalpy,

$$h_{01} = h_{02} \qquad (7.34b)$$

D) Second Law of Thermodynamics

$$\int_{cs} \frac{1}{T} \cancelto{0}{\frac{\dot{Q}}{A}} \, dA = \frac{\partial}{\partial t} \int_{cv} s\rho \cancelto{0}{dV} + \int_{cs} s\rho \vec{V}d\vec{A}$$

or,

$$s_2 - s_1 > 0 \qquad (7.35)$$

where

$$s_2 - s_1 = C_p \ln \frac{T_2}{T_1} - R\ln \frac{P_2}{P_1}$$

E) Equation of State

$$p = \rho RT \qquad (7.36)$$

F) Relationship Between h and T for an Ideal Gas

$$\Delta h = C_p(T_2 - T_1) \qquad (7.26)$$

If the initial state 1 is known, then there is a unique final state 2 for a given initial state.

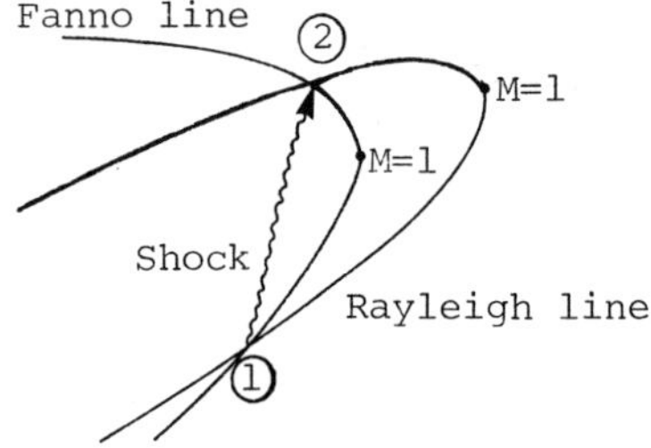

Fig. 7.9 Intersection of Fanno line and Rayleigh line as a solution of the normal shock equations.

The normal shock must satisfy the basic equations plus equation (7.26). A very informative procedure for finding the final state is to make use of the Fanno line and the Rayleigh line. The intersection of the two curves (Fig. 7.9, point 2) represents the conditions of the final state corresponding to that given initial state (point 1).

As a final remark, a normal shock occurs only in a supersonic flow and after the shock there must be a subsonic flow.

The summary of property changes across a normal shock are listed in the following table:

Table 7.3

Property	Effect
Stagnation temperature, T_0	Constant
Entropy, s	Increases
Stagnation pressure, p_0	Decreases
Temperature, T	Increases
Velocity, V	Decreases
Density, ρ	Increases
Pressure, p	Increases
Mach number, M	Decreases

The expressions for the property ratios across the normal shock are the following:

$$\frac{T_2}{T_1} = \frac{1 + \frac{K-1}{2} M_1^2}{1 + \frac{K-1}{2} M_2^2} \tag{7.37}$$

$$\frac{\rho_2}{\rho_1} = \frac{V_1}{V_2} = \frac{M_1}{M_2} \left[\frac{1 + \frac{K-1}{2} M_2^2}{1 + \frac{K-1}{2} M_1^2} \right]^{\frac{1}{2}} \tag{7.38}$$

$$\frac{p_2}{p_1} = \frac{1 + K M_1^2}{1 + K M_2^2} \tag{7.39}$$

$$\frac{p_{02}}{p_{01}} = \frac{p_2}{p_1} \left[\frac{1 + \frac{K-1}{2} M_2^2}{1 + \frac{K-1}{2} M_1^2} \right]^{K/(K-1)} \tag{7.40}$$

The above ratios can be supplied by the use of tables.

7.10 FLOW IN A CONVERGING-DIVERGING NOZZLE

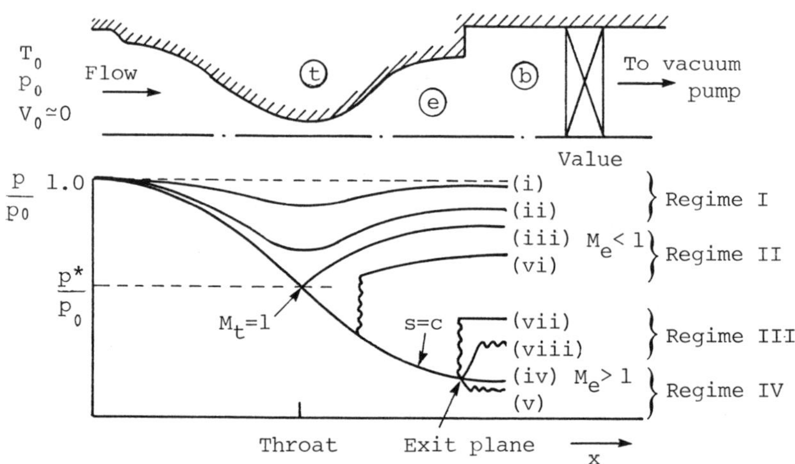

Fig. 7.1Q   Pressure distributions for flow in
a converging-diverging nozzle as a
function of back pressure.

The effect of variations in back pressure, P_b, on the
pressure distribution through a converging-diverging nozzle
is graphically illustrated in Figure 7.10. The flow through
this type of nozzle may be divided into three regimes:

REGIME I : Subsonic Flow

The back pressure, P_b, is equal to the stagnation
pressure, P_o; therefore there is no flow (curve a). As P_b
continues to decrease the flowrate increases but   the   flow
remains subsonic (curves b and c). Deceleration takes place
in the diverging section.

REGIME II : Nozzle Over-Expanding

Here, the pressure at some point in the nozzle is less
than the back pressure ($P_d > P_b > P_f$). The velocity of the

flow is critical, the pressure is minimum, and the flow is choked at the throat. The pressure distribution is represented by curve d. As the pressure is lowered, a shock occurs downstream from the throat and moves until it appears at the exit plane of the nozzle.

REGIME III : Nozzle Under-Expanding

A reduction in the back pressure lower than curve f has no effect on the flow in the nozzle. When $P_b = P_f$, the flow is isentropic through the nozzle and supersonic at the nozzle exit. Nozzles operating at this condition are said to be at design conditions. If the pressure is further lowered a series of oblique expansion waves occur and it cannot be studied using the one-dimensional theory.

7.11 OBLIQUE SHOCK

OBLIQUE SHOCK-RELATIONS

The normal shock is a special case of a more general family of oblique shocks that occur in a supersonic flow. An oblique shock occurs when a supersonic flow changes direction due to a boundary surface converging towards the flow.

Fig. 7.11   Oblique shock wave geometry.

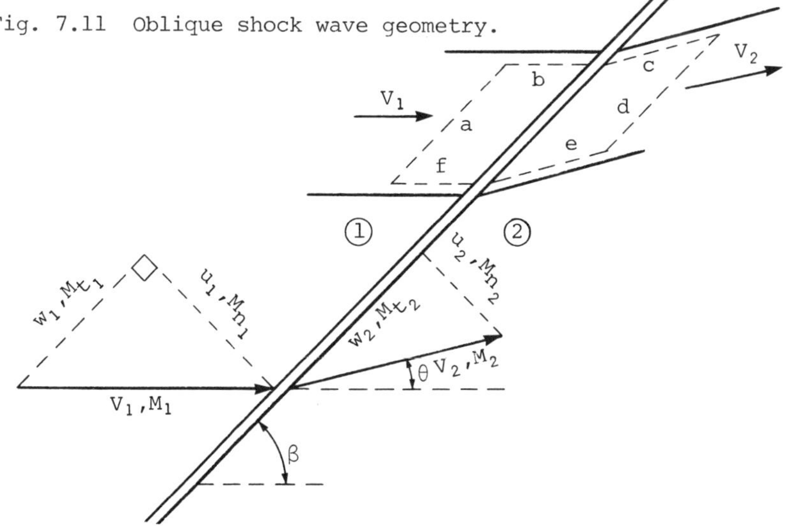

Figure 7.11 illustrates an oblique shock wave in two-dimensional flow.

The basic equations for a steady flow with no body forces applied to the control volume shown in Figure 7.11b are:

A) Continuity Equation

$$0 = \frac{\partial}{\partial t} \int_{cv} \rho \, d\forall + \int_{cs} \rho \vec{V} d\vec{A}$$

or,

$$\rho_1 u_1 = \rho_2 u_2 \tag{7.41}$$

where

ρ_1, ρ_2 = The densities at section 1 and 2, respectively

u_1 = The component of V_1 perpendicular to the shock wave

u_2 = The component of V_2 perpendicular to the shock wave

B) Momentum Equation

$$F_{Sx} + F_{Bx} = \frac{\partial}{\partial t} \int_{cv} V_x \rho \, d\forall + \int_{cs} V_x \rho \vec{v} d\vec{A} \tag{7.42}$$

If we decompose this equation into two components, we have:

a) Parallel to the shock

$$-(\rho_1 u_1)w_1 + (\rho_2 u_2)w_2 = 0$$

or,

$$w_1 = w_2 \tag{7.43}$$

where,

w_1 = The component of V_1 parallel to the shock wave

w_2 = The component of V_2 parallel to the shock wave

b) Perpendicular to the shock

$$-(-\rho_1 u_1)u_1 - (\rho_2 u_2)u_2 = -(p_1 - p_2)$$

$$(7.44)$$

or,

$$P_1 + \rho_1 u_1^2 = p_2 + \rho_2 u_2^2$$

where

$$\rho_1, \rho_2 = \text{The densities at section 1 and 2, respectively}$$

C) First Law of Thermodynmaics

$$\dot{Q} + \dot{W}_s + \dot{W}_{shear} + \dot{W}_{other}$$

$$= \frac{\partial}{\partial t} \int_{cv} e\rho \, d\mathcal{V} + \int_{cv} (e+pv)\rho \vec{V} \cdot d\vec{A}$$

or,

$$-(-p_1 u_1 + p_2 u_2) = -\rho_1 \left(e_1 + \frac{V_1^2}{2}\right) u_1 + \rho_2 \left(e_2 + \frac{V_2^2}{2}\right) u_2$$

or,

$$h_1 + \frac{V_1^2}{2} = h_2 + \frac{V_2^2}{2} \tag{7.45}$$

or,

$$h_1 + \frac{u_1^2}{2} = h_2 + \frac{u_2^2}{2} \tag{7.46}$$

Looking at equations (7.41), (7.44), (7.46) we conclude that the changes across an oblique shock are governed by the normal components of the freestream velocity.

D) Perfect Gas

For an oblique shock wave with $M_{n1} = M_1 \sin b$ and $M_{n2} = M_2 \sin(b-\theta)$, we have for a perfect gas the following relations:

$$\frac{\rho_2}{\rho_1} = \frac{(K+1) M_{n1}^2}{(K-1) M_{n1}^2 + 2} \tag{7.47}$$

$$\frac{p_2}{p_1} = 1 + \frac{2K}{K+1}(M_{n1}^2 - 1) \tag{7.48}$$

$$M_{n_2}^2 = \frac{M_{n_1}^2 + [2/(K-1)]}{[2K/(K-1)]M_{n_1}^2 - 1} \tag{7.49}$$

$$\frac{T_2}{T_1} = \frac{p_2}{p_1} \frac{\rho_1}{\rho_2} \tag{7.50}$$

where

M_{n_1} = Normal component of the upstream Mach number

M_{n_2} = Normal component of the downstream Mach number

K = Specific Heat Ratio

Combining the above equation we obtain the θ-b-M relation which specifies θ as a unique function of M_1 and b, or

$$\tan\theta = 2\cot b \left[\frac{M_1^2 \sin^2 b - 1}{M_1^2(K + \cos 2b) + 2} \right] \tag{7.51}$$

CHAPTER 8

DIMENSIONAL ANALYSIS AND PHYSICAL SIMILARITY

8.1 DIMENSIONAL ANALYSIS, PHYSICAL SIMILARITY

Table 8.1

Quantity	Dimensions	
	(M,L,T)	(F,L,T)
Acceleration	LT^{-2}	LT^{-2}
Area	L^2	L^2
Bulk modulus of elasticity	$ML^{-1}T^{-2}$	FL^{-2}
Density	ML^{-3}	FT^2L^{-4}
Discharge	L^3T^{-1}	L^3T^{-1}
Dynamic viscosity	$ML^{-1}T^{-1}$	$FT\,L^{-2}$
Force	MLT^{-2}	F
Gravity	LT^{-2}	LT^{-2}
Kinematic viscosity	L^2T^{-1}	L^2T^{-1}
Length	L	L
Mass	M	FT^2L^{-1}
Pressure	$ML^{-1}T^{-2}$	FL^{-2}
Specific weight	$ML^{-2}T^{-2}$	FL^{-3}
Tension	MT^{-2}	FL^{-1}
Time	T	T
Velocity	LT^{-1}	LT^{-1}

The study of fluid mechanics is based mainly on experimental results. One technique used to minimize the number of experiments required is dimensional analysis. Dimensional analysis does not provide a complete solution to a problem, but it reveals the mathematical relations among the variables involved. Table 8.1 shows a summary of the quantities, symbols, and dimensions used in fluid mechanics.

Many experiments in fluid mechanics are conducted on scale models rather than on prototypes. In all of these experiments, results taken from tests using the model are applied to the prototype. Physical similarity is a general proposition between the model and the prototype.

8.2 TYPES OF PHYSICAL SIMILARITY

For any comparison between prototype and model to be valid, they must satisfy the following conditions of physical similarity:

A) Geometric Similarity

This physical similarity requires, first, that the model and prototype have the same shape, and second, that their dimensions be related by a constant scale factor.

B) Kinematic Similarity

This physical similarity requires that the velocities be related in magnitude and direction by a constant scale factor.

C) Dynamic Similarity

This physical similarity requires that identical types of forces be related in magnitude and direction by a constant scale factor.

Some of the most important force ratios (dimensionless groups), in dynamic similarity are listed in table 8.2

Table 8.2   Force Ratios

Name of Ratio	Definition	Physical Meaning
Reynolds number, Re	$\dfrac{V \ell \rho}{\mu}$	$\dfrac{\text{Inertia force}}{\text{Viscous force}}$
Froude number, Fr	$\dfrac{V}{(\ell g)^{\frac{1}{2}}}$	$\dfrac{\text{Inertia force}}{\text{Gravity force}}$
Weber number, We	$V \left[\dfrac{\ell \rho}{\gamma} \right]^{\frac{1}{2}}$	$\dfrac{\text{Inertia force}}{\text{Surface tension force}}$
Mach number, M	$\dfrac{V}{a}$	$\dfrac{\text{Inertia force}}{\text{Elastic force}}$
Euler number, Eu	$\dfrac{\Delta p}{\frac{1}{2} \rho V^2}$	$\dfrac{\text{Pressure force}}{\text{Inertia force}}$

8.3 THE BUCKINGHAM'S Π THEOREM

This theorem states that the number of independent dimensionless groups used to describe a problem in which there are n variables and m dimensions is equal to n m. Mathematically it can be represented as

$$f(\pi_1, \pi_2, \ldots, \pi_{n-m}) = 0$$

where $\quad \pi \equiv$ The dimensionless group.

8.4 DIMENSIONAL ANALYSIS OF A PROBLEM

The dimensional analysis of a problem is performed in the following steps:

A) A list of parameters is selected.

B) The dimensionless π parameters are obtained by using the π theorem.

Six steps are used to find the dimensionless parameters:

1) List all the variables involved.

2) Select a set of primary dimensions.

3) List the dimensions of all variables in terms of primary dimensions.

4) Select from the list of variables, a number of repeating variables equal to the number of primary dimensions, m, including all the primary dimensions.

5) Write the π parameters in terms of the unknown exponents and write the equations so that the sum of exponents be zero. Solve the equation to obtain n-m dimensionless group.

6) Check if each group obtained is dimensionless.

C) The relation among π parameters is determined experimentally.

HANDBOOK OF MATHEMATICAL, SCIENTIFIC, and ENGINEERING

FORMULAS, TABLES, FUNCTIONS, GRAPHS, TRANSFORMS

A particularly useful reference for those in math, science, engineering and other technical fields. Includes the most-often used formulas, tables, transforms, functions, and graphs which are needed as tools in solving problems. The entire field of special functions is also covered. A large amount of scientific data which is often of interest to scientists and engineers has been included.

Available at your local bookstore or order directly from us by sending in coupon below.

RESEARCH and EDUCATION ASSOCIATION
61 Ethel Road W., Piscataway, New Jersey 08854
Phone: (201) 819-8880

☐ Payment enclosed

☐ Visa
☐ Master Card

Please check one box:

Charge Card Number

Expiration Date _______ / _______
Mo   Yr

Please ship the "Math Handbook" @ $24.85 plus $4.00 for shipping.

Name ____

Address ____

City _______________________ State __________ Zip __________

THE PROBLEM SOLVERS

The "PROBLEM SOLVERS" are comprehensive supplemental textbooks designed to save time in finding solutions to problems. Each "PROBLEM SOLVER" is the of its kind ever produced in its field. It is the product of a massive effort to illustrate al any imaginable problem in exceptional depth, detail, and clarity. Each problem is wo out in detail with step-by-step solution, and the problems are arranged in order of compl from elementary to advanced. Each book is fully indexed for locating problems rapid

ADVANCED CALCULUS
ALGEBRA & TRIGONOMETRY
AUTOMATIC CONTROL
   SYSTEMS/ROBOTICS
BIOLOGY
BUSINESS, MANAGEMENT,
   & FINANCE
CALCULUS
CHEMISTRY
COMPLEX VARIABLES
COMPUTER SCIENCE
DIFFERENTIAL EQUATIONS
ECONOMICS
ELECTRICAL MACHINES
ELECTRIC CIRCUITS
ELECTROMAGNETICS
ELECTRONIC COMMUNICATIONS
ELECTRONICS
FINITE & DISCRETE MATH
FLUID MECHANICS/DYNAMICS
GENETICS

GEOMETRY:
PLANE · SOLID · ANALYTIC
HEAT TRANSFER
LINEAR ALGEBRA
MACHINE DESIGN
MECHANICS : STATICS · DYNAMICS
NUMERICAL ANALYSIS
OPERATIONS RESEARCH
OPTICS
ORGANIC CHEMISTRY
PHYSICAL CHEMISTRY
PHYSICS
PRE-CALCULUS
PSYCHOLOGY
STATISTICS
STRENGTH OF MATERIALS &
   MECHANICS OF SOLIDS
TECHNICAL DESIGN GRAPHIC
THERMODYNAMICS
TRANSPORT PHENOMENA :
MOMENTUM · ENERGY · MASS
VECTOR ANALYSIS

If you would like more information about any of these books, complete the cou below and return it to us or go to your local bookstore.

RESEARCH and EDUCATION ASSOCIATION
61 Ethel Road W. • Piscataway • New Jersey 08854
Phone: (201) 819-8880

Please send me more information about your Problem Solver Books

Name ____

Address ____

City _____________________________ State __________ Zip __________